Air Quality Monitoring and Control Strategies

Air Quality Monitoring and Control Strategies

Edited by
Bertram O'Neal

Larsen & Keller
www.larsen-keller.com

Air Quality Monitoring and Control Strategies
Edited by Bertram O'Neal
ISBN: 978-1-63549-022-0 (Hardback)

⊟ Larsen & Keller

Published by Larsen and Keller Education,
5 Penn Plaza,
19th Floor,
New York, NY 10001, USA

Cataloging-in-Publication Data

Air quality monitoring and control strategies / edited by Bertram O'Neal.
    p. cm.
Includes bibliographical references and index.
ISBN 978-1-63549-022-0
1. Air quality. 2. Air--Pollution--Measurement.
3. Air quality management. 4. Environmental monitoring. I. O'Neal, Bertram.
TD883 .A37 2017
628.53--dc23

The publisher's policy is to use permanent paper from mills that operate a sustainable forestry policy. Furthermore, the publisher ensures that the text paper and cover boards used have met acceptable environmental accreditation standards.

Printed and bound in the United States of America.

For more information regarding Larsen and Keller Education and its products, please visit the publisher's website www.larsen-keller.com

Table of Contents

Preface

This book elucidates the concepts and innovative models around prospective developments with respect to air quality monitoring and control strategies. It provides great detail about the fundamental concepts of this subject. Air pollution is a growing concern for our world today. Air quality control and monitoring has been necessary for increasing the standard of life and in order to achieve this, certain laws, rules and regulations have been passed by countries and organizations in various capacities. This text is a compilation of chapters that discuss the most vital concepts of this field. Most of the topics introduced in this book covers new devices and technologies related to air quality monitoring. Different approaches, evaluations and methodologies have been included in it. It will serve as a valuable source of reference for students.

To facilitate a deeper understanding of the contents of this book a short introduction of every chapter is written below:

Chapter 1- Air pollution is the existence of substances into the Earth's atmosphere. It causes diseases, allergies and death to humans. The major source of air pollution comes from anthropogenic sources. This chapter will provide an integrated understanding of air quality monitoring.

Chapter 2- Air pollution dispersion is the amount of pollution that is distributed into the atmosphere. Air pollution comes from anthropogenic or natural sources. The aspects elucidated are atmospheric dispersion modeling, roadway air dispersion modeling, ADMS 3, CALPUFF and NAME. The section is an overview of the major aspects of air pollution dispersion.

Chapter 3- Indoor air quality or IAQ refers to the quality of the air within buildings. The pollution that is caused indoors is a major health hazard in a number of countries; one of the major causes of air pollution is the burning of biomass. Passive smoking, HVAC, air conditioning and air handle are some of the reasons for varied indoor air quality. The major categories of indoor air quality are dealt with great details in the chapter.

Chapter 4- Cyclonic separation is the method of removing particulates; these particulates are removed from air, gas and also from liquid streams. The alternative technologies used are selective catalytic reduction, exhaust gas recirculation, biofilter, thermal oxidizer and vapor recovery. This section serves as a source to understand the main technologies used in controlling air pollution.

Chapter 5- An electrostatic precipitator is a device that is used to filter air; it helps in removing particles of smoke and dust. The alternative devices that are used in controlling air pollution are dust collectors, scrubbers, catalytic converters and air purge systems. The topics discussed in the chapter are of great importance to broaden the existing knowledge on air pollution.

Chapter 6- The two main methods explained are geothermal heat pumps and seasonal thermal energy storage. Geothermal heat pump transfers heat from the ground and is used as a cooling or heating system whereas seasonal thermal energy storage is used as storage for a number of months. Air quality improvement is best understood in confluence with the major topics listed in the following section.

Chapter 7- Air quality law is the law that governs the release of air pollutants into the air. It helps in the regulation of the quality of air inside and around building. The aspects elucidates in this chapter are of vital importance and provides a better understanding of air quality law.

I owe the completion of this book to the never-ending support of my family, who supported me throughout the project.

Editor

Understanding Air Quality Monitoring

Air pollution is the existence of substances into the Earth's atmosphere. It causes diseases, allergies and death to humans. The major source of air pollution comes from anthropogenic sources. This chapter will provide an integrated understanding of air quality monitoring.

Air Pollution

Air pollution is the introduction of particulates, biological molecules, and many harmful substances into Earth's atmosphere, causing diseases, allergies, death to humans, damage to other living organisms such as animals and food crops, or the natural or built environment. Air pollution may come from anthropogenic or natural sources.

Air pollution from a fossil fuel power station

The atmosphere is a complex natural gaseous system that is essential to support life on planet Earth.

Indoor air pollution and urban air quality are listed as two of the world's worst toxic pollution problems in the 2008 Blacksmith Institute World's Worst Polluted Places report. According to the 2014 WHO report, air pollution in 2012 caused the deaths of around 7 million people worldwide, an estimate roughly matched by the International Energy Agency.

Pollutants

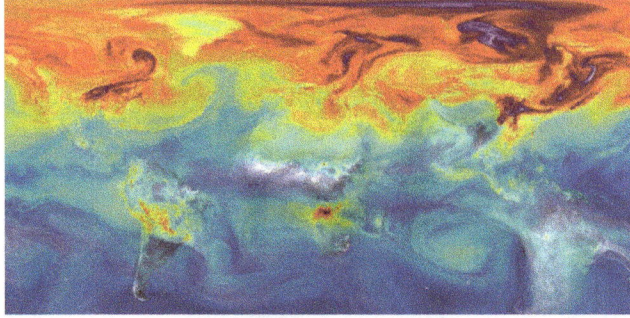

Carbon dioxide in Earth's atmosphere if *half* of global-warming
emissions are *not* absorbed. (NASA simulation; 9 November 2015)

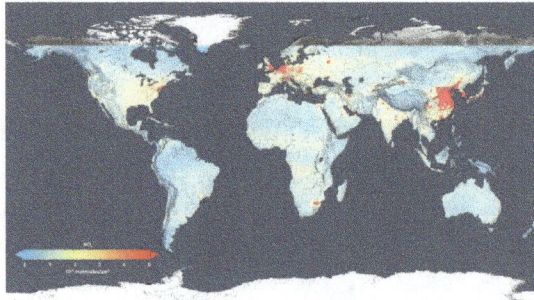

Nitrogen dioxide 2014 - global air quality levels (released 14 December 2015).

An air pollutant is a substance in the air that can have adverse effects on humans and the ecosystem. The substance can be solid particles, liquid droplets, or gases. A pollutant can be of natural origin or man-made. Pollutants are classified as primary or secondary. Primary pollutants are usually produced from a process, such as ash from a volcanic eruption. Other examples include carbon monoxide gas from motor vehicle exhaust, or the sulfur dioxide released from factories. Secondary pollutants are not emitted directly. Rather, they form in the air when primary pollutants react or interact. Ground level ozone is a prominent example of a secondary pollutant. Some pollutants may be both primary and secondary: they are both emitted directly and formed from other primary pollutants.

Before flue-gas desulfurization was installed, the emissions from
this power plant in New Mexico contained excessive amounts of sulfur dioxide.

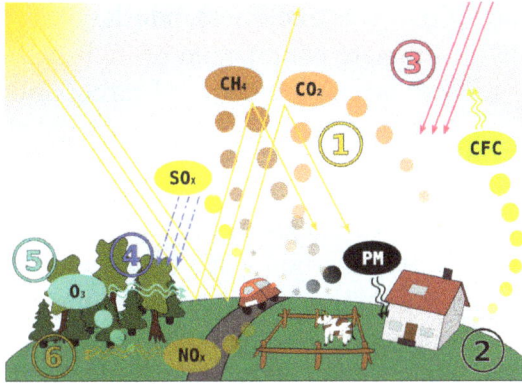

Schematic drawing, causes and effects of air pollution: (1) greenhouse effect,
(2) particulate contamination, (3) increased UV radiation, (4) acid rain,
(5) increased ground level ozone concentration, (6) increased levels of nitrogen oxides.

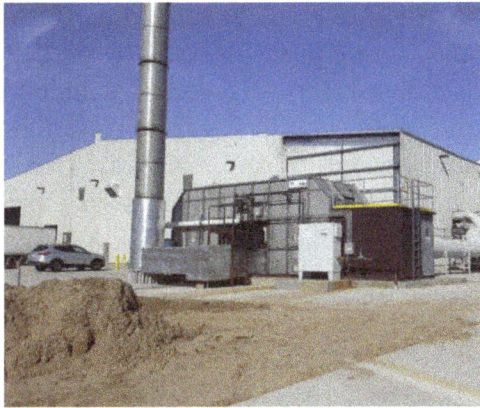

Thermal oxidizers are air pollution abatement options for hazardous
air pollutants (HAPs), volatile organic compounds (VOCs), and odorous emissions.

Major primary pollutants produced by human activity include:

- Sulfur oxides (SO_x) - particularly sulfur dioxide, a chemical compound with the formula SO_2. SO_2 is produced by volcanoes and in various industrial processes. Coal and petroleum often contain sulfur compounds, and their combustion generates sulfur dioxide. Further oxidation of SO_2, usually in the presence of a catalyst such as NO_2, forms H_2SO_4, and thus acid rain. This is one of the causes for concern over the environmental impact of the use of these fuels as power sources.

- Nitrogen oxides (NO_x) - Nitrogen oxides, particularly nitrogen dioxide, are expelled from high temperature combustion, and are also produced during thunderstorms by electric discharge. They can be seen as a brown haze dome above or a plume downwind of cities. Nitrogen dioxide is a chemical compound with the formula NO_2. It is one of several nitrogen oxides. One of the most prominent air pollutants, this reddish-brown toxic gas has a characteristic sharp, biting odor.

- Carbon monoxide (CO) - CO is a colorless, odorless, toxic yet non-irritating gas. It is a product by incomplete combustion of fuel such as natural gas, coal or wood. Vehicular exhaust is a major source of carbon monoxide.

- Volatile organic compounds (VOC) - VOCs are a well-known outdoor air pollutant. They are categorized as either methane (CH_4) or non-methane (NMVOCs). Methane is an extremely efficient greenhouse gas which contributes to enhanced global warming. Other hydrocarbon VOCs are also significant greenhouse gases because of their role in creating ozone and prolonging the life of methane in the atmosphere. This effect varies depending on local air quality. The aromatic NMVOCs benzene, toluene and xylene are suspected carcinogens and may lead to leukemia with prolonged exposure. 1,3-butadiene is another dangerous compound often associated with industrial use.

- Particulates, alternatively referred to as particulate matter (PM), atmospheric particulate matter, or fine particles, are tiny particles of solid or liquid suspended in a gas. In contrast, aerosol refers to combined particles and gas. Some particulates occur naturally, originating from volcanoes, dust storms, forest and grassland fires, living vegetation, and sea spray. Human activities, such as the burning of fossil fuels in vehicles, power plants and various industrial processes also generate significant amounts of aerosols. Averaged worldwide, anthropogenic aerosols—those made by human activities—currently account for approximately 10 percent of our atmosphere. Increased levels of fine particles in the air are linked to health hazards such as heart disease, altered lung function and lung cancer.

- Persistent free radicals connected to airborne fine particles are linked to cardiopulmonary disease.

- Toxic metals, such as lead and mercury, especially their compounds.

- Chlorofluorocarbons (CFCs) - harmful to the ozone layer; emitted from products are currently banned from use. These are gases which are released from air conditioners, refrigerators, aerosol sprays, etc. CFC's on being released into the air rises to stratosphere. Here they come in contact with other gases and damage the ozone layer. This allows harmful ultraviolet rays to reach the earth's surface. This can lead to skin cancer, disease to eye and can even cause damage to plants.

- Ammonia (NH_3) - emitted from agricultural processes. Ammonia is a compound with the formula NH_3. It is normally encountered as a gas with a characteristic pungent odor. Ammonia contributes significantly to the nutritional needs of terrestrial organisms by serving as a precursor to foodstuffs and fertilizers. Ammonia, either directly or indirectly, is also a building block for the synthesis of many pharmaceuticals. Although in wide use, ammonia is both caustic and hazardous. In the atmosphere, ammonia reacts with oxides of nitrogen and sulfur to form secondary particles.

- Odours — such as from garbage, sewage, and industrial processes

- Radioactive pollutants - produced by nuclear explosions, nuclear events, war explosives, and natural processes such as the radioactive decay of radon.

Secondary pollutants include:

- Particulates created from gaseous primary pollutants and compounds in photochemical smog. Smog is a kind of air pollution. Classic smog results from large amounts of coal burning in an area caused by a mixture of smoke and sulfur dioxide. Modern smog does not usually come from coal but from vehicular and industrial emissions that are acted on in the atmosphere by ultraviolet light from the sun to form secondary pollutants that also combine with the primary emissions to form photochemical smog.

- Ground level ozone (O_3) formed from NO_x and VOCs. Ozone (O_3) is a key constituent of the troposphere. It is also an important constituent of certain regions of the stratosphere commonly known as the Ozone layer. Photochemical and chemical reactions involving it drive many of the chemical processes that occur in the atmosphere by day and by night. At abnormally high concentrations brought about by human activities (largely the combustion of fossil fuel), it is a pollutant, and a constituent of smog.

- Peroxyacetyl nitrate (PAN) - similarly formed from NO_x and VOCs.

Minor air pollutants include:

- A large number of minor hazardous air pollutants. Some of these are regulated in USA under the Clean Air Act and in Europe under the Air Framework Directive

- A variety of persistent organic pollutants, which can attach to particulates

Persistent organic pollutants (POPs) are organic compounds that are resistant to environmental degradation through chemical, biological, and photolytic processes. Because of this, they have been observed to persist in the environment, to be capable of long-range transport, bioaccumulate in human and animal tissue, biomagnify in food chains, and to have potentially significant impacts on human health and the environment.

Sources

There are various locations, activities or factors which are responsible for releasing pollutants into the atmosphere. These sources can be classified into two major categories.

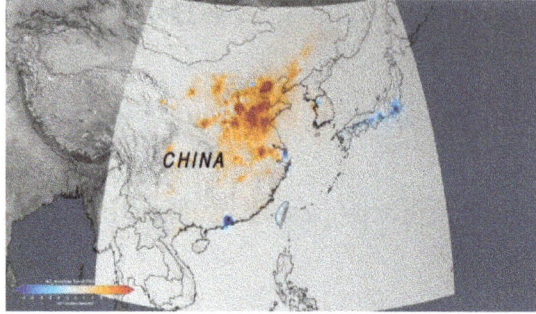

This video provides an overview of a NASA study on the human fingerprint on global air quality.

Anthropogenic (Man-made) Sources:

Controlled burning of a field outside of Statesboro, Georgia in preparation for spring planting.

These are mostly related to the burning of multiple types of fuel.

- Stationary sources include smoke stacks of power plants, manufacturing facilities (factories) and waste incinerators, as well as furnaces and other types of fuel-burning heating devices. In developing and poor countries, traditional biomass burning is the major source of air pollutants; traditional biomass includes wood, crop waste and dung.

- Mobile sources include motor vehicles, marine vessels, and aircraft.

- Controlled burn practices in agriculture and forest management. Controlled or prescribed burning is a technique sometimes used in forest management, farming, prairie restoration or greenhouse gas abatement. Fire is a natural part of both forest and grassland ecology and controlled fire can be a tool for foresters. Controlled burning stimulates the germination of some desirable forest trees, thus renewing the forest.

- Fumes from paint, hair spray, varnish, aerosol sprays and other solvents

- Waste deposition in landfills, which generate methane. Methane is highly flammable and may form explosive mixtures with air. Methane is also an asphyxiant and may displace oxygen in an enclosed space. Asphyxia or suffocation may result if the oxygen concentration is reduced to below 19.5% by displacement.

- Military resources, such as nuclear weapons, toxic gases, germ warfare and rocketry

Natural Sources:

Dust storm approaching Stratford, Texas.

- Dust from natural sources, usually large areas of land with little or no vegetation

- Methane, emitted by the digestion of food by animals, for example cattle

- Radon gas from radioactive decay within the Earth's crust. Radon is a colorless, odorless, naturally occurring, radioactive noble gas that is formed from the decay of radium. It is considered to be a health hazard. Radon gas from natural sources can accumulate in buildings, especially in confined areas such as the basement and it is the second most frequent cause of lung cancer, after cigarette smoking.

- Smoke and carbon monoxide from wildfires

- Vegetation, in some regions, emits environmentally significant amounts of Volatile organic compounds (VOCs) on warmer days. These VOCs react with primary anthropogenic pollutants—specifically, NO_x, SO_2, and anthropogenic organic carbon compounds — to produce a seasonal haze of secondary pollutants. Black gum, poplar, oak and willow are some examples of vegetation that can produce abundant VOCs. The VOC production from these species result in ozone levels up to eight times higher than the low-impact tree species.

- Volcanic activity, which produces sulfur, chlorine, and ash particulates

Emission Factors

Air pollutant emission factors are reported representative values that attempt to relate the quantity of a pollutant released to the ambient air with an activity associated with the release of that pollutant. These factors are usually expressed as the weight of pollutant divided by a unit weight, volume, distance, or duration of the activity emitting the pollutant (e.g., kilograms of particulate emitted per tonne of coal burned). Such factors facilitate estimation of emissions from various sources of air pollution. In most cases, these factors are simply averages of all available data of acceptable quality, and are generally assumed to be representative of long-term averages.

Beijing air on a 2005-day after rain (left) and a smoggy day (right)

There are 12 compounds in the list of persistent organic pollutants. Dioxins and furans are two of them and intentionally created by combustion of organics, like open burning of plastics. These compounds are also endocrine disruptors and can mutate the human genes.

The United States Environmental Protection Agency has published a compilation of air pollutant emission factors for a wide range of industrial sources. The United Kingdom, Australia, Canada and many other countries have published similar compilations, as well as the European Environment Agency.

Exposure

Air pollution risk is a function of the hazard of the pollutant and the exposure to that pollutant. Air pollution exposure can be expressed for an individual, for certain groups (e.g. neighborhoods or children living in a country), or for entire populations. For example, one may want to calculate the exposure to a hazardous air pollutant for a geographic area, which includes the various microenvironments and age groups. This can be calculated as an inhalation exposure. This would account for daily exposure in various settings (e.g. different indoor micro-environments and outdoor locations). The exposure needs to include different age and other demographic groups, especially infants, children, pregnant women and other sensitive subpopulations. The exposure to an air pollutant must integrate the concentrations of the air pollutant with respect to the time spent in each setting and the respective inhalation rates for each subgroup for each specific time that the subgroup is in the setting and engaged in particular activities (playing, cooking, reading, working, etc.). For example, a small child's inhalation rate will be less than that of an adult. A child engaged in vigorous exercise will have a higher respiration rate than the same child in a sedentary activity. The daily exposure, then, needs to reflect the time spent in each micro-environmental setting and the type of activities in these settings. The air pollutant concentration in each microactivity/microenvironmental setting is summed to indicate the exposure.

Indoor Air Quality (IAQ)

Air quality monitoring, New Delhi, India.

A lack of ventilation indoors concentrates air pollution where people often spend the majority of their time. Radon (Rn) gas, a carcinogen, is exuded from the Earth in certain locations and trapped inside houses. Building materials including carpeting and plywood emit formaldehyde (H_2CO) gas. Paint and solvents give off volatile organic compounds (VOCs) as they dry. Lead paint can degenerate into dust and be inhaled. Intentional air pollution is introduced with the use of air fresheners, incense, and other scented items. Controlled wood fires in stoves and fireplaces can add significant amounts of smoke particulates into the air, inside and out. Indoor pollution fatalities may be caused by using pesticides and other chemical sprays indoors without proper ventilation.

Carbon monoxide poisoning and fatalities are often caused by faulty vents and chimneys, or by the burning of charcoal indoors or in a confined space, such as a tent. Chronic carbon monoxide poisoning can result even from poorly-adjusted pilot lights. Traps are built into all domestic plumbing to keep sewer gas and hydrogen sulfide, out of interiors. Clothing emits tetrachloroethylene, or other dry cleaning fluids, for days after dry cleaning.

Though its use has now been banned in many countries, the extensive use of asbestos in industrial and domestic environments in the past has left a potentially very dangerous material in many localities. Asbestosis is a chronic inflammatory medical condition affecting the tissue of the lungs. It occurs after long-term, heavy exposure to asbestos from asbestos-containing materials in structures. Sufferers have severe dyspnea (shortness of breath) and are at an increased risk regarding several different types of lung cancer. As clear explanations are not always stressed in non-technical literature, care should be taken to distinguish between several forms of relevant diseases. According to the World Health Organisation (WHO), these may defined as; asbestosis, *lung cancer*, and *Peritoneal Mesothelioma* (generally a very rare form of cancer, when more widespread it is almost always associated with prolonged exposure to asbestos).

Biological sources of air pollution are also found indoors, as gases and airborne particulates. Pets produce dander, people produce dust from minute skin flakes and de-composed hair, dust mites in bedding, carpeting and furniture produce enzymes and micrometre-sized fecal droppings, inhabitants emit methane, mold forms on walls and generates mycotoxins and spores, air conditioning systems can incubate Legionnaires' disease and mold, and houseplants, soil and surrounding gardens can produce pollen, dust, and mold. Indoors, the lack of air circulation allows these airborne pollutants to accumulate more than they would otherwise occur in nature.

Health Effects

Air pollution is a significant risk factor for a number of pollution-related diseases and health conditions including respiratory infections, heart disease, COPD, stroke and lung cancer. The health effects caused by air pollution may include difficulty in breath-ing, wheezing, coughing, asthma and worsening of existing respiratory and cardiac conditions. These effects can result in increased medication use, increased doctor or emergency room visits, more hospital admissions and premature death. The human health effects of poor air quality are far reaching, but principally affect the body's re-spiratory system and the cardiovascular system. Individual reactions to air pollutants depend on the type of pollutant a person is exposed to, the degree of exposure, and the individual's health status and genetics. The most common sources of air pollution include particulates, ozone, nitrogen dioxide, and sulphur dioxide. Children aged less than five years that live in developing countries are the most vulnerable population in terms of total deaths attributable to indoor and outdoor air pollution.

Mortality

The World Health Organization estimated in 2014 that every year air pollution causes the premature death of some 7 million people worldwide. India has the highest death rate due to air pollution. India also has more deaths from asthma than any other na-tion according to the World Health Organization. In December 2013 air pollution was estimated to kill 500,000 people in China each year. There is a positive correlation between pneumonia-related deaths and air pollution from motor vehicle emissions.

Annual premature European deaths caused by air pollution are estimated at 430,000. An important cause of these deaths is nitrogen dioxide and other nitrogen oxides (NOx) emitted by road vehicles. Across the European Union, air pollution is estimated to re-duce life expectancy by almost nine months. Causes of deaths include strokes, heart disease, COPD, lung cancer, and lung infections.

Urban outdoor air pollution is estimated to cause 1.3 million deaths worldwide per year. Children are particularly at risk due to the immaturity of their respiratory organ systems.

The US EPA estimates that a proposed set of changes in diesel engine technology (*Tier 2*) could result in 12,000 fewer *premature mortalities*, 15,000 fewer heart attacks,

6,000 fewer emergency room visits by children with asthma, and 8,900 fewer respiratory-related hospital admissions each year in the United States.

The US EPA has estimated that limiting ground-level ozone concentration to 65 parts per billion, would avert 1,700 to 5,100 premature deaths nationwide in 2020 compared with the 75-ppb standard. The agency projected the more protective standard would also prevent an additional 26,000 cases of aggravated asthma, and more than a million cases of missed work or school. Following this assessment, the EPA acted to protect public health by lowering the National Ambient Air Quality Standards (NAAQS) for ground-level ozone to 70 parts per billion (ppb).

A new economic study of the health impacts and associated costs of air pollution in the Los Angeles Basin and San Joaquin Valley of Southern California shows that more than 3,800 people die prematurely (approximately 14 years earlier than normal) each year because air pollution levels violate federal standards. The number of annual premature deaths is considerably higher than the fatalities related to auto collisions in the same area, which average fewer than 2,000 per year.

Diesel exhaust (DE) is a major contributor to combustion-derived particulate matter air pollution. In several human experimental studies, using a well-validated exposure chamber setup, DE has been linked to acute vascular dysfunction and increased thrombus formation.

The mechanisms linking air pollution to increased cardiovascular mortality are uncertain, but probably include pulmonary and systemic inflammation.

Cardiovascular Disease

A 2007 review of evidence found ambient air pollution exposure is a risk factor correlating with increased total mortality from cardiovascular events (range: 12% to 14% per 10 microg/m³ increase).

Air pollution is also emerging as a risk factor for stroke, particularly in developing countries where pollutant levels are highest. A 2007 study found that in women, air pollution is not associated with hemorrhagic but with ischemic stroke. Air pollution was also found to be associated with increased incidence and mortality from coronary stroke in a cohort study in 2011. Associations are believed to be causal and effects may be mediated by vasoconstriction, low-grade inflammation and atherosclerosis Other mechanisms such as autonomic nervous system imbalance have also been suggested.

Lung Disease

Chronic obstructive pulmonary disease (COPD) includes diseases such as chronic bronchitis and emphysema.

Research has demonstrated increased risk of developing asthma and COPD from increased exposure to traffic-related air pollution. Additionally, air pollution has been associated with increased hospitalization and mortality from asthma and COPD.

A study conducted in 1960-1961 in the wake of the Great Smog of 1952 compared 293 London residents with 477 residents of Gloucester, Peterborough, and Norwich, three towns with low reported death rates from chronic bronchitis. All subjects were male postal truck drivers aged 40 to 59. Compared to the subjects from the outlying towns, the London subjects exhibited more severe respiratory symptoms (including cough, phlegm, and dyspnea), reduced lung function (FEV_1 and peak flow rate), and increased sputum production and purulence. The differences were more pronounced for subjects aged 50 to 59. The study controlled for age and smoking habits, so concluded that air pollution was the most likely cause of the observed differences.

It is believed that much like cystic fibrosis, by living in a more urban environment serious health hazards become more apparent. Studies have shown that in urban areas patients suffer mucus hypersecretion, lower levels of lung function, and more self-diagnosis of chronic bronchitis and emphysema.

Cancer

Cancer mainly the result of environmental factors.

A review of evidence regarding whether ambient air pollution exposure is a risk factor for cancer in 2007 found solid data to conclude that long-term exposure to PM2.5 (fine particulates) increases the overall risk of non-accidental mortality by 6% per a 10 microg/m³ increase. Exposure to PM2.5 was also associated with an increased risk of mortality from lung cancer (range: 15% to 21% per 10 microg/m³ increase) and total cardiovascular mortality (range: 12% to 14% per a 10 microg/m³ increase). The review further noted that living close to busy traffic appears to be associated with elevated risks of these three outcomes --- increase in lung cancer deaths, cardiovascular deaths, and overall non-accidental deaths. The reviewers also found suggestive evidence that exposure to PM2.5 is positively associated with mortality from coronary heart diseases and exposure to SO_2 increases mortality from lung cancer, but the data was insufficient to provide solid conclusions. Another investigation showed that higher activity level increases deposition fraction of aerosol particles in human lung and recommended avoiding heavy activities like running in outdoor space at polluted areas.

In 2011, a large Danish epidemiological study found an increased risk of lung cancer for patients who lived in areas with high nitrogen oxide concentrations. In this study, the association was higher for non-smokers than smokers. An additional Danish study, also in 2011, likewise noted evidence of possible associations between air pollution and other forms of cancer, including cervical cancer and brain cancer.

In December 2015, medical scientists reported that cancer is overwhelmingly a result of environmental factors, and not largely down to bad luck. Maintaining a healthy weight, eating a healthy diet, minimizing alcohol and eliminating smoking reduces the risk of developing the disease, according to the researchers.

Children

In the United States, despite the passage of the Clean Air Act in 1970, in 2002 at least 146 million Americans were living in non-attainment areas—regions in which the concentration of certain air pollutants exceeded federal standards. These dangerous pollutants are known as the criteria pollutants, and include ozone, particulate matter, sulfur dioxide, nitrogen dioxide, carbon monoxide, and lead. Protective measures to ensure children's health are being taken in cities such as New Delhi, India where buses now use compressed natural gas to help eliminate the "pea-soup" smog. A recent study in Europe has found that exposure to ultrafine particles can increase blood pressure in children.

"Clean" Areas

Even in the areas with relatively low levels of air pollution, public health effects can be significant and costly, since a large number of people breathe in such pollutants. A 2005 scientific study for the British Columbia Lung Association showed that a small improvement in air quality (1% reduction of ambient PM2.5 and ozone concentrations) would produce $29 million in annual savings in the Metro Vancouver region in 2010. This finding is based on health valuation of lethal (death) and sub-lethal (illness) affects.

Central Nervous System

Data is accumulating that air pollution exposure also affects the central nervous system.

In a June 2014 study conducted by researchers at the University of Rochester Medical

Center, published in the journal Environmental Health Perspectives, it was discovered that early exposure to air pollution causes the same damaging changes in the brain as autism and schizophrenia. The study also shows that air pollution also affected short-term memory, learning ability, and impulsivity. Lead researcher Professor Deborah Cory-Slechta said that "When we looked closely at the ventricles, we could see that the white matter that normally surrounds them hadn't fully developed. It appears that inflammation had damaged those brain cells and prevented that region of the brain from developing, and the ventricles simply expanded to fill the space. Our findings add to the growing body of evidence that air pollution may play a role in autism, as well as in other neurodevelopmental disorders." Air pollution has a more significant negative effect on males than on females.

In 2015, experimental studies reported the detection of significant episodic (situational) cognitive impairment from impurities in indoor air breathed by test subjects who were not informed about changes in the air quality. Researchers at the Harvard University and SUNY Upstate Medical University and Syracuse University measured the cognitive performance of 24 participants in three different controlled laboratory atmospheres that simulated those found in "conventional" and "green" buildings, as well as green buildings with enhanced ventilation. Performance was evaluated objectively using the widely used Strategic Management Simulation software simulation tool, which is a well-validated assessment test for executive decision-making in an unconstrained situation allowing initiative and improvisation. Significant deficits were observed in the performance scores achieved in increasing concentrations of either volatile organic compounds (VOCs) or carbon dioxide, while keeping other factors constant. The highest impurity levels reached are not uncommon in some classroom or office environments.

Agricultural Effects

In India in 2014, it was reported that air pollution by black carbon and ground level ozone had cut crop yields in the most affected areas by almost half in 2010 when compared to 1980 levels.

Economic Effects

Air pollution costs the world economy $5 trillion per year as a result of productivity losses and degraded quality of life, according to a joint study by the World Bank and the Institute for Health Metrics and Evaluation (IHME) at the University of Washington These productivity losses are caused by deaths due to diseases caused by air pollution. One out of ten deaths in 2013 was caused by diseases associated with air pollution and the problem is getting worse. The problem is even more acute in the developing world. "Children under age 5 in lower-income countries are more than 60 times as likely to die from exposure to air pollution as children in high-income countries." The report states that additional economic losses caused by air pollution, including health costs and the adverse effect on agricultural and other productivity were not calculated in the report, and thus the actual costs to the world economy are far higher than $5 trillion.

Historical Disasters

The world's worst short-term civilian pollution crisis was the 1984 Bhopal Disaster in India. Leaked industrial vapours from the Union Carbide factory, belonging to Union Carbide, Inc., U.S.A. (later bought by Dow Chemical Company), killed at least 3787 people and injured anywhere from 150,000 to 600,000. The United Kingdom suffered its worst air pollution event when the December 4 Great Smog of 1952 formed over London. In six days more than 4,000 died and more recent estimates put the figure at nearer 12,000. An accidental leak of anthrax spores from a biological warfare laboratory in the former USSR in 1979 near Sverdlovsk is believed to have caused at least 64 deaths. The worst single incident of air pollution to occur in the US occurred in Donora, Pennsylvania in late October, 1948, when 20 people died and over 7,000 were injured.

Alternatives to Pollution

There are now practical alternatives to the three principal causes of air pollution.

- Combustion of fossil fuels for space heating can be replaced by using ground source heat pumps and seasonal thermal energy storage.

- Electric power generation from burning fossil fuels can be replaced by power generation from nuclear and renewables.

- Motor vehicles driven by fossil fuels, a key factor in urban air pollution, can be replaced by electric vehicles.

Reduction Efforts

There are various air pollution control technologies and strategies available to reduce air pollution. At its most basic level, land-use planning is likely to involve zoning and transport infrastructure planning. In most developed countries, land-use planning is an important part of social policy, ensuring that land is used efficiently for the benefit of the wider economy and population, as well as to protect the environment.

Because a large share of air pollution is caused by combustion of fossil fuels such as coal and oil, the reduction of these fuels can reduce air pollution drastically. Most effective is the switch to clean power sources such as wind power, solar power, hydro power which don't cause air pollution. Efforts to reduce pollution from mobile sources includes primary regulation (many developing countries have permissive regulations), expanding regulation to new sources (such as cruise and transport ships, farm equipment, and small gas-powered equipment such as string trimmers, chainsaws, and snowmobiles), increased fuel efficiency (such as through the use of hybrid vehicles), conversion to cleaner fuels or conversion to electric vehicles.

Titanium dioxide has been researched for its ability to reduce air pollution. Ultraviolet light will release free electrons from material, thereby creating free radicals, which break up VOCs and NOx gases. One form is superhydrophilic.

In 2014, Prof. Tony Ryan and Prof. Simon Armitage of University of Sheffield prepared a 10 meter by 20 meter-sized poster coated with microscopic, pollution-eating nanoparticles of titanium dioxide. Placed on a building, this giant poster can absorb the toxic emission from around 20 cars each day.

A very effective means to reduce air pollution is the transition to renewable energy. According to a study published in Energy and Environmental Science in 2015 the switch to 100% renewable energy in the United States would eliminate about 62,000 premature mortalities per year and about 42,000 in 2050, if no biomass were used. This would save about $600 billion in health costs a year due to reduced air pollution in 2050, or about 3.6% of the 2014 U.S. gross domestic product.

Control Devices

The following items are commonly used as pollution control devices in industry and transportation. They can either destroy contaminants or remove them from an exhaust stream before it is emitted into the atmosphere.

- Particulate control

    o Mechanical collectors (dust cyclones, multicyclones)

    o Electrostatic precipitators An electrostatic precipitator (ESP), or electrostatic air cleaner is a particulate collection device that removes particles from a flowing gas (such as air), using the force of an induced electrostatic charge. Electrostatic precipitators are highly efficient filtration devices that minimally impede the flow of gases through the device, and can easily remove fine particulates such as dust and smoke from the air stream.

    o Baghouses Designed to handle heavy dust loads, a dust collector consists of a blower, dust filter, a filter-cleaning system, and a dust receptacle or dust removal system (distinguished from air cleaners which utilize disposable filters to remove the dust).

    o Particulate scrubbers Wet scrubber is a form of pollution control technology. The term describes a variety of devices that use pollutants from a furnace flue gas or from other gas streams. In a wet scrubber, the polluted gas stream is brought into contact with the scrubbing liquid, by spraying it with the liquid, by forcing it through a pool of liquid, or by some other contact method, so as to remove the pollutants.

- Scrubbers
  - Baffle spray scrubber
  - Cyclonic spray scrubber
  - Ejector venturi scrubber
  - Mechanically aided scrubber
  - Spray tower
  - Wet scrubber
- NOx control
  - Low NOx burners
  - Selective catalytic reduction (SCR)
  - Selective non-catalytic reduction (SNCR)
  - NOx scrubbers
  - Exhaust gas recirculation
  - Catalytic converter (also for VOC control)
- VOC abatement
  - Adsorption systems, using activated carbon, such as Fluidized Bed Concentrator
  - Flares
  - Thermal oxidizers
  - Catalytic converters
  - Biofilters
  - Absorption (scrubbing)
  - Cryogenic condensers
  - Vapor recovery systems

- Acid Gas/SO$_2$ control
    - o Wet scrubbers
    - o Dry scrubbers
    - o Flue-gas desulfurization
- Mercury control
    - o Sorbent Injection Technology
    - o Electro-Catalytic Oxidation (ECO)
    - o K-Fuel
- Dioxin and furan control
- Miscellaneous associated equipment
    - o Source capturing systems
    - o Continuous emissions monitoring systems (CEMS)

Regulations

Smog in Cairo

In general, there are two types of air quality standards. The first class of standards (such as the U.S. National Ambient Air Quality Standards and E.U. Air Quality Directive) set maximum atmospheric concentrations for specific pollutants. Environmental agencies enact regulations which are intended to result in attainment of these target levels. The second class (such as the North American Air Quality Index) take the form of a scale with various thresholds, which is used to communicate to the public the relative risk of outdoor activity. The scale may or may not distinguish between different pollutants.

Canada

In Canada, air pollution and associated health risks are measured with the Air Quality

Health Index or (AQHI). It is a health protection tool used to make decisions to reduce short-term exposure to air pollution by adjusting activity levels during increased levels of air pollution.

The Air Quality Health Index or "AQHI" is a federal program jointly coordinated by Health Canada and Environment Canada. However, the AQHI program would not be possible without the commitment and support of the provinces, municipalities and NGOs. From air quality monitoring to health risk communication and community engagement, local partners are responsible for the vast majority of work related to AQHI implementation. The AQHI provides a number from 1 to 10+ to indicate the level of health risk associated with local air quality. Occasionally, when the amount of air pollution is abnormally high, the number may exceed 10. The AQHI provides a local air quality current value as well as a local air quality maximums forecast for today, tonight and tomorrow and provides associated health advice.

1	2	3	4	5	6	7	8	9	10	+

Risk:	Low (1-3)	Moderate (4-6)	High (7-10)	Very high (above 10)

As it is now known that even low levels of air pollution can trigger discomfort for the sensitive population, the index has been developed as a continuum: The higher the number, the greater the health risk and need to take precautions. The index describes the level of health risk associated with this number as 'low', 'moderate', 'high' or 'very high', and suggests steps that can be taken to reduce exposure.

Health Risk	Air Quality Health Index	Health Messages	
		At Risk population	General Population
Low	1-3	Enjoy your usual outdoor activities.	Ideal air quality for outdoor activities
Moderate	4-6	Consider reducing or rescheduling strenuous activities outdoors if you are experiencing symptoms.	No need to modify your usual outdoor activities unless you experience symptoms such as coughing and throat irritation.
High	7-10	Reduce or reschedule strenuous activities outdoors. Children and the elderly should also take it easy.	Consider reducing or rescheduling strenuous activities outdoors if you experience symptoms such as coughing and throat irritation.
Very high	Above 10	Avoid strenuous activities outdoors. Children and the elderly should also avoid outdoor physical exertion and should stay indoors.	Reduce or reschedule strenuous activities outdoors, especially if you experience symptoms such as coughing and throat irritation.

The measurement is based on the observed relationship of Nitrogen Dioxide (NO_2), ground-level Ozone (O_3) and particulates ($PM_{2.5}$) with mortality, from an analysis of several Canadian cities. Significantly, all three of these pollutants can pose health risks, even at low levels of exposure, especially among those with pre-existing health problems.

When developing the AQHI, Health Canada's original analysis of health effects included five major air pollutants: particulates, ozone, and nitrogen dioxide (NO2), as well as sulfur dioxide (SO_2), and carbon monoxide (CO). The latter two pollutants provided little information in predicting health effects and were removed from the AQHI formulation.

The AQHI does not measure the effects of odour, pollen, dust, heat or humidity.

Germany

TA Luft is the German air quality regulation.

Hotspots

Air pollution hotspots are areas where air pollution emissions expose individuals to increased negative health effects. They are particularly common in highly populated, urban areas, where there may be a combination of stationary sources (e.g. industrial facilities) and mobile sources (e.g. cars and trucks) of pollution. Emissions from these sources can cause respiratory disease, childhood asthma, cancer, and other health problems. Fine particulate matter such as diesel soot, which contributes to more than 3.2 million premature deaths around the world each year, is a significant problem. It is very small and can lodge itself within the lungs and enter the bloodstream. Diesel soot is concentrated in densely populated areas, and one in six people in the U.S. live near a diesel pollution hot spot.

While air pollution hotspots affect a variety of populations, some groups are more likely to be located in hotspots. Previous studies have shown disparities in exposure to pollution by race and/or income. Hazardous land uses (toxic storage and disposal facilities, manufacturing facilities, major roadways) tend to be located where property values and income levels are low. Low socioeconomic status can be a proxy for other kinds of social vulnerability, including race, a lack of ability to influence regulation and a lack of ability to move to neighborhoods with less environmental pollution. These communities bear a disproportionate burden of environmental pollution and are more likely to face health risks such as cancer or asthma.

Studies show that patterns in race and income disparities not only indicate a higher exposure to pollution but also higher risk of adverse health outcomes. Communities characterized by low socioeconomic status and racial minorities can be more vulnerable to cumulative adverse health impacts resulting from elevated exposure to pollutants than more privileged communities. Blacks and Latinos generally face more pollution than whites and Asians, and low-income communities bear a higher burden of risk than affluent ones. Racial discrepancies are particularly distinct in suburban areas of the US South and metropolitan areas of the US West. Residents in public housing, who are generally low-income and cannot move to healthier neighborhoods, are highly affected by nearby refineries and chemical plants.

Cities

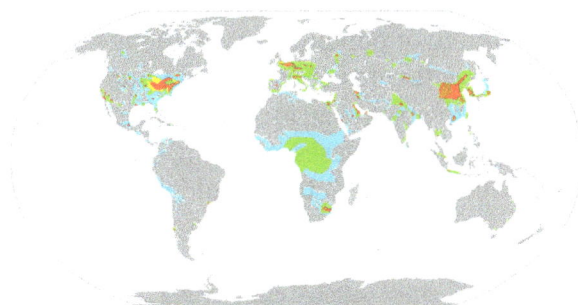

Nitrogen dioxide concentrations as measured from satellite 2002-2004

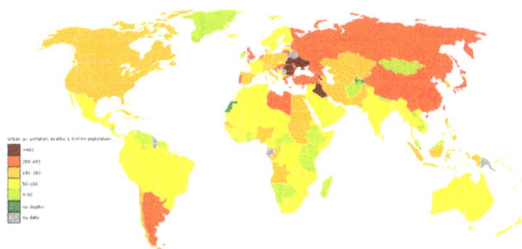

Deaths from air pollution in 2004

Air pollution is usually concentrated in densely populated metropolitan areas, especially in developing countries where environmental regulations are relatively lax or nonexistent. However, even populated areas in developed countries attain unhealthy levels of pollution, with Los Angeles and Rome being two examples. Between 2002 and 2011 the incidence of lung cancer in Beijing near doubled. While smoking remains the leading cause of lung cancer in China, the number of smokers is falling while lung cancer rates are rising. Another project focusing on the effects on pollution in vegetation has been researched by the local university in Sheffield, UK.

National-scale Air Toxics Assessments 1995-2005

The national-scale air toxics assessment(NATA) is an evaluation of air toxics by the U.S. EPA. EPA has furnished four assessments that characterize nationwide chronic cancer risk estimates and noncancer hazards from inhaling air toxics. The lates was from 2005, and made publicly available in early 2011.

"EPA developed the NATA as a state-of-the-science screening tool for State/Local/ Tribal Agencies to prioritize pollutants, emission sources and locations of interest for further study, in order to gain a better understanding of the risks. NATA assessments do not incorporate refined information about emission sources, but rather, use general information about sources to develop estimates of risks which are more likely to overestimate impacts than underestimate them. NATA provides estimates of the risk of cancer and other serious health effects from breathing (inhaling) air toxics in order to inform both national and more localized efforts to identify and prioritize air toxics, emission source types and locations which are of greatest potential concern in terms of contributing to population risk. This in turn helps air pollution experts focus limited analytical resources on areas and or populations where the potential for health risks are highest. Assessments include estimates of cancer and non-cancer health effects based on chronic exposure from outdoor sources, including assessments of non-cancer health effects for Diesel Particulate Matter. Assessments provide a snapshot of the outdoor air quality and the risks to human health that would result if air toxic emissions levels remained unchanged."

Most polluted cities by PM	
Particulate matter, µg/m³ (2004)	City
168	Cairo, Egypt
150	Delhi, India
128	Kolkata, India (Calcutta)
125	Tianjin, China
123	Chongqing, China
109	Kanpur, India
109	Lucknow, India
104	Jakarta, Indonesia
101	Shenyang, China

Governing Urban air Pollution

In Europe, Council Directive 96/62/EC on ambient air quality assessment and management provides a common strategy against which member states can "set objectives for ambient air quality in order to avoid, prevent or reduce harmful effects on human health and the environment . . . and improve air quality where it is unsatisfactory".

On 25 July 2008 in the case Dieter Janecek v Freistaat Bayern CURIA, the European Court of Justice ruled that under this directive citizens have the right to require national authorities to implement a short term action plan that aims to maintain or achieve compliance to air quality limit values.

This important case law appears to confirm the role of the EC as centralised regulator to European nation-states as regards air pollution control. It places a supranational legal obligation on the UK to protect its citizens from dangerous levels of air pollution, furthermore superseding national interests with those of the citizen.

In 2010, the European Commission (EC) threatened the UK with legal action against the successive breaching of PM10 limit values. The UK government has identified that if fines are imposed, they could cost the nation upwards of £300 million per year.

In March 2011, the Greater London Built-up Area remains the only UK region in breach of the EC's limit values, and has been given 3 months to implement an emergency action plan aimed at meeting the EU Air Quality Directive. The City of London has dangerous levels of PM10 concentrations, estimated to cause 3000 deaths per year within the city. As well as the threat of EU fines, in 2010 it was threatened with legal action for scrapping the western congestion charge zone, which is claimed to have led to an increase in air pollution levels.

In response to these charges, Boris Johnson, Mayor of London, has criticised the current need for European cities to communicate with Europe through their nation state's central government, arguing that in future "A great city like London" should be permitted to bypass its government and deal directly with the European Commission regarding its air quality action plan.

This can be interpreted as recognition that cities can transcend the traditional national government organisational hierarchy and develop solutions to air pollution using global governance networks, for example through transnational relations. Transnational relations include but are not exclusive to national governments and intergovernmental organisations, allowing sub-national actors including cities and regions to partake in air pollution control as independent actors.

Particularly promising at present are global city partnerships. These can be built into networks, for example the C40 Cities Climate Leadership Group, of which London is a member. The C40 is a public 'non-state' network of the world's leading cities that aims to curb their greenhouse emissions. The C40 has been identified as 'governance from the middle' and is an alternative to intergovernmental policy. It has the potential to improve urban air quality as participating cities "exchange information, learn from best practices and consequently mitigate carbon dioxide emissions independently from national government decisions". A criticism of the C40 network is that its exclusive nature limits influence to participating cities and risks drawing resources away from less powerful city and regional actors.

Atmospheric Dispersion

The basic technology for analyzing air pollution is through the use of a variety of mathematical models for predicting the transport of air pollutants in the lower atmosphere. The principal methodologies are:

- Point source dispersion, used for industrial sources

- Line source dispersion, used for airport and roadway air dispersion modeling

- Area source dispersion, used for forest fires or duststorms

- Photochemical models, used to analyze reactive pollutants that form smog

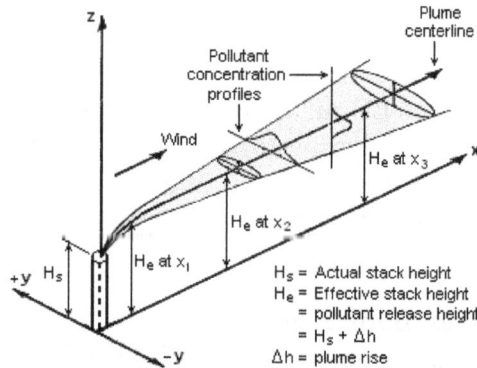

Visualization of a buoyant Gaussian air pollution dispersion
plume as used in many atmospheric dispersion models.

The point source problem is the best understood, since it involves simpler mathematics and has been studied for a long period of time, dating back to about the year 1900. It uses a Gaussian dispersion model for continuous buoyant pollution plumes to predict the air pollution isopleths, with consideration given to wind velocity, stack height, emission rate and stability class (a measure of atmospheric turbulence). This model has been extensively validated and calibrated with experimental data for all sorts of atmospheric conditions.

The roadway air dispersion model was developed starting in the late 1950s and early 1960s in response to requirements of the National Environmental Policy Act and the U.S. Department of Transportation (then known as the Federal Highway Administration) to understand impacts of proposed new highways upon air quality, especially in urban areas. Several research groups were active in this model development, among which were: the Environmental Research and Technology (ERT) group in Lexington, Massachusetts, the ESL Inc. group in Sunnyvale, California and the California Air Resources Board group in Sacramento, California. The research of the ESL group received a boost with a contract award from the United States Environmental Protection Agency to validate a line source model using sulfur hexafluoride as a tracer gas. This program was successful in validating the line source model developed by ESL Inc. Some of the earliest uses of the model were in court cases involving highway air pollution; the Arlington, Virginia portion of Interstate 66 and the New Jersey Turnpike widening project through East Brunswick, New Jersey.

Area source models were developed in 1971 through 1974 by the ERT and ESL groups, but addressed a smaller fraction of total air pollution emissions, so that their use and need was not as widespread as the line source model, which enjoyed hundreds of different applications as early as the 1970s. Similarly photochemical models were developed primarily in the 1960s and 70s, but their use was more specialized and for regional needs, such as understanding smog formation in Los Angeles, California.

Air Quality Index

An air quality index (AQI) is a number used by government agencies to communicate to the public how polluted the air currently is or how polluted it is forecast to become. As the AQI increases, an increasingly large percentage of the population is likely to experience increasingly severe adverse health effects. Different countries have their own air quality indices, corresponding to different national air quality standards. Some of these are the Air Quality Health Index (Canada), the Air Pollution Index (Malaysia), and the Pollutant Standards Index (Singapore).

Smog builds up under an inversion in Almaty, Kazakhstan resulting in a high AQI

Wildfires give rise to an elevated AQI in parts of Greece

Definition and Usage

Computation of the AQI requires an air pollutant concentration over a specified averaging period, obtained from an air monitor or model. Taken together, concentration and time represent the dose of the air pollutant. Health effects corresponding to a given dose are established by epidemiological research. Air pollutants vary in potency, and the function used to convert from air pollutant concentration to AQI varies by pollutant. Air quality index values are typically grouped into ranges. Each range is assigned a descriptor, a color code, and a standardized public health advisory.

An air quality measurement station in Edinburgh, Scotland

The AQI can increase due to an increase of air emissions (for example, during rush hour traffic or when there is an upwind forest fire) or from a lack of dilution of air pollutants. Stagnant air, often caused by an anticyclone, temperature inversion, or low wind speeds lets air pollution remain in a local area, leading to high concentrations of pollutants, chemical reactions between air contaminants and hazy conditions.

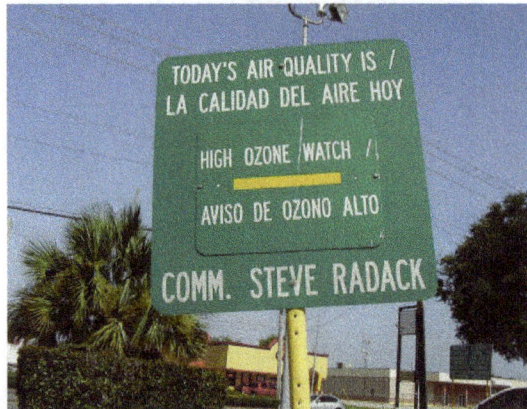

Signboard in Gulfton, Houston indicating an ozone watch

On a day when the AQI is predicted to be elevated due to fine particle pollution, an agency or public health organization might:

- advise sensitive groups, such as the elderly, children, and those with respiratory or cardiovascular problems to avoid outdoor exertion.

- declare an "action day" to encourage voluntary measures to reduce air emissions, such as using public transportation.

- recommend the use of masks to keep fine particles from entering the lungs

During a period of very poor air quality, such as an air pollution episode, when the AQI indicates that acute exposure may cause significant harm to the public health, agencies may invoke emergency plans that allow them to order major emitters (such as coal burning industries) to curtail emissions until the hazardous conditions abate.

Most air contaminants do not have an associated AQI. Many countries monitor ground-level ozone, particulates, sulfur dioxide, carbon monoxide and nitrogen dioxide, and calculate air quality indices for these pollutants.

The definition of the AQI in a particular nation reflects the discourse surrounding the development of national air quality standards in that nation. A website allowing government agencies anywhere in the world to submit their real-time air monitoring data for display using a common definition of the air quality index has recently become available.

Indices by Location

Canada

Air quality in Canada has been reported for many years with provincial Air Quality Indices (AQIs). Significantly, AQI values reflect air quality management objectives, which are based on the lowest achievable emissions rate, and not exclusively concern for human health. The Air Quality Health Index or (AQHI) is a scale designed to help understand the impact of air quality on health. It is a health protection tool used to make decisions to reduce short-term exposure to air pollution by adjusting activity levels during increased levels of air pollution. The Air Quality Health Index also provides advice on how to improve air quality by proposing behavioural change to reduce the environmental footprint. This index pays particular attention to people who are sensitive to air pollution. It provides them with advice on how to protect their health during air quality levels associated with low, moderate, high and very high health risks.

The Air Quality Health Index provides a number from 1 to 10+ to indicate the level of health risk associated with local air quality. On occasion, when the amount of air pollution is abnormally high, the number may exceed 10. The AQHI provides a local air quality current value as well as a local air quality maximums forecast for today, tonight, and tomorrow, and provides associated health advice.

| 1 | 2 | 3 | 4 | 5 | 6 | 7 | 8 | 9 | 10 | + |

| Risk: | Low (1–3) | Moderate (4–6) | High (7–10) | Very high (above 10) |

Health Risk	Air Quality Health Index	Health Messages	
		At Risk population	*General Population
Low	1–3	Enjoy your usual outdoor activities.	Ideal air quality for outdoor activities
Moderate	4–6	Consider reducing or rescheduling strenuous activities outdoors if you are experiencing symptoms.	No need to modify your usual outdoor activities unless you experience symptoms such as coughing and throat irritation.
High	7–10	Reduce or reschedule strenuous activities outdoors. Children and the elderly should also take it easy.	Consider reducing or rescheduling strenuous activities outdoors if you experience symptoms such as coughing and throat irritation.
Very high	Above 10	Avoid strenuous activities outdoors. Children and the elderly should also avoid outdoor physical exertion.	Reduce or reschedule strenuous activities outdoors, especially if you experience symptoms such as coughing and throat irritation.

Hong Kong

On the 30th December 2013 Hong Kong replaced the Air Pollution Index with a new index called the *Air Quality Health Index*. This index is on a scale of 1 to 10+ and considers four air pollutants: ozone; nitrogen dioxide; sulphur dioxide and particulate matter (including PM10 and PM2.5). For any given hour the AQHI is calculated from the sum of the percentage excess risk of daily hospital admissions attributable to the 3-hour moving average concentrations of these four pollutants. The AQHIs are grouped into five AQHI health risk categories with health advice provided:

Health risk category	AQHI
Low	1
Low	2
Low	3
Medium	4
Medium	5
Medium	6
High	7
Very High	8
Very High	9
Very High	10
Serious	10+

Each of the health risk categories has advice with it. At the *low* and *moderate* levels the public are advised that they can continue normal activities. For the *high* category, children, the elderly and people with heart or respiratory illnesses are advising to reduce outdoor physical exertion. Above this (*very high* or *serious*) the general public are also advised to reduce or avoid outdoor physical exertion.

Mainland China

China's Ministry of Environmental Protection (MEP) is responsible for measuring the level of air pollution in China. As of 1 January 2013, MEP monitors daily pollution level in 163 of its major cities. The API level is based on the level of 6 atmospheric pollutants, namely sulfur dioxide (SO_2), nitrogen dioxide (NO_2), suspended particulates smaller than 10 μm in aerodynamic diameter (PM_{10}), suspended particulates smaller than 2.5 μm in aerodynamic diameter ($PM_{2.5}$),   carbon monoxide (CO), and ozone (O_3) measured at the monitoring stations throughout each city.

AQI Mechanics An individual score (IAQI) is assigned to the level of each pollutant and the final AQI is the highest of those 6 scores. The pollutants can be measured quite differently. $PM_{2.5}$, PM_{10} concentration are measured as average per 24h. SO_2, NO_2, O_3, CO are measured as average per hour. The final API value is calculated per hour according to a formula published by the MEP.

The scale for each pollutant is non-linear, as is the final AQI score. Thus an AQI of 100 does not mean twice the pollution of AQI at 50, nor does it mean twice as harmful. While an AQI of 50 from day 1 to 182 and AQI of 100 from day 183 to 365 does provide an annual average of 75, it does *not* mean the pollution is acceptable even if the benchmark of 100 is deemed safe. This is because the benchmark is a 24-hour target. The annual average must match against the annual target. It is entirely possible to have safe air every day of the year but still fail the annual pollution benchmark.

AQI and Health Implications (HJ 663-2012)

AQI	Air Pollution Level	Health Implications
0–50	Excellent	No health implications.
51–100	Good	Few hypersensitive individuals should reduce outdoor exercise.
101–150	Lightly Polluted	Slight irritations may occur, individuals with breathing or heart problems should reduce outdoor exercise.
151–200	Moderately Polluted	Slight irritations may occur, individuals with breathing or heart problems should reduce outdoor exercise.
201–300	Heavily Polluted	Healthy people will be noticeably affected. People with breathing or heart problems will experience reduced endurance in activities. These individuals and elders should remain indoors and restrict activities.
300+	Severely Polluted	Healthy people will experience reduced endurance in activities. There may be strong irritations and symptoms and may trigger other illnesses. Elders and the sick should remain indoors and avoid exercise. Healthy individuals should avoid outdoor activities.

India

The Minister for Environment, Forests & Climate Change Shri Prakash Javadekar launched The National Air Quality Index (AQI) in New Delhi on 17 September 2014 under the Swachh Bharat Abhiyan. It is outlined as 'One Number- One Colour-One Description' for the common man to judge the air quality within his vicinity. The index constitutes part of the Government's mission to introduce the culture of cleanliness. Institutional and infrastructural measures are being undertaken in order to ensure that the mandate of cleanliness is fulfilled across the country and the Ministry of Environment, Forests & Climate Change proposed to discuss the issues concerned regarding quality of air with the Ministry of Human Resource Development in order to include this issue as part of the sensitisation programme in the course curriculum.

While the earlier measuring index was limited to three indicators, the current measurement index had been made quite comprehensive by the addition of five additional parameters. Under the current measurement of air quality there are 8 parameters . The initiatives undertaken by the Ministry recently aimed at balancing environment and conservation and development as air pollution has been a matter of environmental and health concerns, particularly in urban areas.

The Central Pollution Control Board along with State Pollution Control Boards has been operating National Air Monitoring Program (NAMP) covering 240 cities of the country having more than 342 monitoring stations. In addition, continuous monitoring systems that provide data on near real-time basis are also installed in a few cities. They provide information on air quality in public domain in simple linguistic terms that is easily understood by a common person. Air Quality Index (AQI) is one such tool for effective dissemination of air quality information to people. As such an Expert Group comprising medical professionals, air quality experts, academia, advocacy groups, and SPCBs was constituted and a technical study was awarded to IIT Kanpur. IIT Kanpur and the Expert Group recommended an AQI scheme in 2014.

There are six AQI categories, namely Good, Satisfactory, Moderately polluted, Poor, Very Poor, and Severe. The proposed AQI will consider eight pollutants (PM_{10}, $PM_{2.5}$, NO_2, SO_2, CO, O_3, NH_3, and Pb) for which short-term (up to 24-hourly averaging period) National Ambient Air Quality Standards are prescribed. Based on the measured ambient concentrations, corresponding standards and likely health impact, a sub-index is calculated for each of these pollutants. The worst sub-index reflects overall AQI. Associated likely health impacts for different AQI categories and pollutants have been also been suggested, with primary inputs from the medical expert members of the group. The AQI values and corresponding ambient concentrations (health breakpoints) as well as associated likely health impacts for the identified eight pollutants are as follows:

AQI Category, Pollutants and Health Breakpoints								
AQI Category (Range)	PM$_{10}$ (24hr)	PM$_{2.5}$ (24hr)	NO$_2$ (24hr)	O$_3$ (8hr)	CO (8hr)	SO$_2$ (24hr)	NH$_3$ (24hr)	Pb (24hr)
Good (0-50)	0-50	0-30	0-40	0-50	0-1.0	0-40	0-200	0-0.5
Satisfactory (51-100)	51-100	31-60	41-80	51-100	1.1-2.0	41-80	201-400	0.5-1.0
Moderately polluted (101-200)	101-250	61-90	81-180	101-168	2.1-10	81-380	401-800	1.1-2.0
Poor (201-300)	251-350	91-120	181-280	169-208	10-17	381-800	801-1200	2.1-3.0
Very poor (301-400)	351-430	121-250	281-400	209-748	17-34	801-1600	1200-1800	3.1-3.5
Severe (401-500)	430+	250+	400+	748+	34+	1600+	1800+	3.5+

AQI	Associated Health Impacts
Good (0-50)	Minimal impact
Satisfactory (51-100)	May cause minor breathing discomfort to sensitive people.
Moderately polluted (101–200)	May cause breathing discomfort to people with lung disease such as asthma, and discomfort to people with heart disease, children and older adults.
Poor (201-300)	May cause breathing discomfort to people on prolonged exposure, and discomfort to people with heart disease.
Very poor (301-400)	May cause respiratory illness to the people on prolonged exposure. Effect may be more pronounced in people with lung and heart diseases.
Severe (401-500)	May cause respiratory impact even on healthy people, and serious health impacts on people with lung/heart disease. The health impacts may be experienced even during light physical activity.

Mexico

The air quality in Mexico City is reported in IMECAs. The IMECA is calculated using the measurements of average times of the chemicals ozone (O$_3$), sulphur dioxide (SO$_2$), nitrogen dioxide (NO$_2$), carbon monoxide (CO), particles smaller than 2.5 micrometers (PM$_{2.5}$), and particles smaller than 10 micrometers (PM$_{10}$).

Singapore

Singapore uses the Pollutant Standards Index to report on its air quality, with details of the calculation similar but not identical to that used in Malaysia and Hong Kong The PSI chart below is grouped by index values and descriptors, according to the National Environment Agency.

PSI	Descriptor	General Health Effects
0–50		None
51–100	Moderate	Few or none for the general population

101–200	Unhealthy	Mild aggravation of symptoms among susceptible persons i.e. those with underlying conditions such as chronic heart or lung ailments; transient symptoms of irritation e.g. eye irritation, sneezing or coughing in some of the healthy population.
201–300	Very Unhealthy	Moderate aggravation of symptoms and decreased tolerance in persons with heart or lung disease; more widespread symptoms of transient irritation in the healthy population.
301–400	Hazardous	Early onset of certain diseases in addition to significant aggravation of symptoms in susceptible persons; and decreased exercise tolerance in healthy persons.
Above 400	Hazardous	PSI levels above 400 may be life-threatening to ill and elderly persons. Healthy people may experience adverse symptoms that affect normal activity.

South Korea

The Ministry of Environment of South Korea uses the Comprehensive Air-quality Index (CAI) to describe the ambient air quality based on the health risks of air pollution. The index aims to help the public easily understand the air quality and protect people's health. The CAI is on a scale from 0 to 500, which is divided into six categories. The higher the CAI value, the greater the level of air pollution. Of values of the five air pollutants, the highest is the CAI value. The index also has associated health effects and a colour representation of the categories as shown below.

CAI	Description	Health Implications
0–50	Good	A level that will not impact patients suffering from diseases related to air pollution.
51–100	Moderate	A level that may have a meager impact on patients in case of chronic exposure.
101–150	Unhealthy for sensitive groups	A level that may have harmful impacts on patients and members of sensitive groups.
151–250	Unhealthy	A level that may have harmful impacts on patients and members of sensitive groups (children, aged or weak people), and also cause the general public unpleasant feelings.
251–500	Very unhealthy	A level that may have a serious impact on patients and members of sensitive groups in case of acute exposure.

The N Seoul Tower on Namsan Mountain in central Seoul, South Korea, is illuminated in blue, from sunset to 23:00 and 22:00 in winter, on days where the air quality in Seoul is 45 or less. During the spring of 2012, the Tower was lit up for 52 days, which is four days more than in 2011.

United Kingdom

The most commonly used air quality index in the UK is the *Daily Air Quality Index* recommended by the Committee on Medical Effects of Air Pollutants (COMEAP). This index has ten points, which are further grouped into 4 bands: low, moderate, high and very high. Each of the bands comes with advice for at-risk groups and the general population.

Air pollution banding	Value	Health messages for At-risk individuals	Health messages for General population
Low	1–3	Enjoy your usual outdoor activities.	Enjoy your usual outdoor activities.
Moderate	4–6	Adults and children with lung problems, and adults with heart problems, who experience symptoms, should consider reducing strenuous physical activity, particularly outdoors.	Enjoy your usual outdoor activities.
High	7–9	Adults and children with lung problems, and adults with heart problems, should reduce strenuous physical exertion, particularly outdoors, and particularly if they experience symptoms. People with asthma may find they need to use their reliever inhaler more often. Older people should also reduce physical exertion.	Anyone experiencing discomfort such as sore eyes, cough or sore throat should consider reducing activity, particularly outdoors.
Very High	10	Adults and children with lung problems, adults with heart problems, and older people, should avoid strenuous physical activity. People with asthma may find they need to use their reliever inhaler more often.	Reduce physical exertion, particularly outdoors, especially if you experience symptoms such as cough or sore throat.

The index is based on the concentrations of 5 pollutants. The index is calculated from the concentrations of the following pollutants: Ozone, Nitrogen Dioxide, Sulphur Dioxide, PM2.5 (particles with an aerodynamic diameter less than 2.5 µm) and PM10. The breakpoints between index values are defined for each pollutant separately and the overall index is defined as the maximum value of the index. Different averaging periods are used for different pollutants.

Index	Ozone, Running 8 hourly mean ($\mu g/m^3$)	Nitrogen Dioxide, Hourly mean ($\mu g/m^3$)	Sulphur Dioxide, 15 minute mean ($\mu g/m^3$)	PM2.5 Particles, 24 hour mean ($\mu g/m^3$)	PM10 Particles, 24 hour mean ($\mu g/m^3$)
1	0-33	0-67	0-88	0-11	0-16
2	34-66	68-134	89-177	12-23	17-33
3	67-100	135-200	178-266	24-35	34-50
4	101-120	201-267	267-354	36-41	51-58
5	121-140	268-334	355-443	42-47	59-66

6	141-160	335-400	444-532	48-53	67-75
7	161-187	401-467	533-710	54-58	76-83
8	188-213	468-534	711-887	59-64	84-91
9	214-240	535-600	888-1064	65-70	92-100
10	≥ 241	≥ 601	≥ 1065	≥ 71	≥ 101

Europe

To present the air quality situation in European cities in a comparable and easily under-standable way, all detailed measurements are transformed into a single relative figure: the Common Air Quality Index (or CAQI) Three different indices have been developed by Citeair to enable the comparison of three different time scale:.

- An hourly index, which describes the air quality today, based on hourly values and updated every hours,

- A daily index, which stands for the general air quality situation of yesterday, based on daily values and updated once a day,

- An annual index, which represents the city's general air quality conditions through-out the year and compare to European air quality norms. This index is based on the pollutants year average compare to annual limit values, and updated once a year.

However, the proposed indices and the supporting common web site www.airquali-tynow.eu are designed to give a dynamic picture of the air quality situation in each city but not for compliance checking.

The Hourly and Daily Common Indices

These indices have 5 levels using a scale from 0 (very low) to > 100 (very high), it is a relative measure of the amount of air pollution. They are based on 3 pollutants of major concern in Europe: PM10, NO2, O3 and will be able to take into account to 3 additional pollutants (CO, PM2.5 and SO2) where data are also available.

The calculation of the index is based on a review of a number of existing air quality indi-ces, and it reflects EU alert threshold levels or daily limit values as much as possible. In order to make cities more comparable, independent of the nature of their monitoring network two situations are defined:

- Background, representing the general situation of the given agglomeration (based on urban background monitoring sites),

- Roadside, being representative of city streets with a lot of traffic, (based on roadside monitoring stations)

The indices values are updated hourly (for those cities that supply hourly data) and yesterdays daily indices are presented.

Common Air Quality Index Legend:

Pollution	Index Value
Very low	0/25
Low	25/50
Medium	50/75
High	75/100
Very high	>100

The Common Annual Air Quality Index

The common annual air quality index provides a general overview of the air quality situation in a given city all the year through and regarding to the European norms.

It is also calculated both for background and traffic conditions but its principle of calculation is different from the hourly and daily indices. It is presented as a distance to a target index, this target being derived from the EU directives (annual air quality standards and objectives):

- If the index is higher than 1: for one or more pollutants the limit values are not met.

- If the index is below 1: on average the limit values are met.

The annual index is aimed at better taking into account long term exposure to air pollution based on distance to the target set by the EU annual norms, those norms being linked most of the time to recommendations and health protection set up by World Health Organisation.

United States

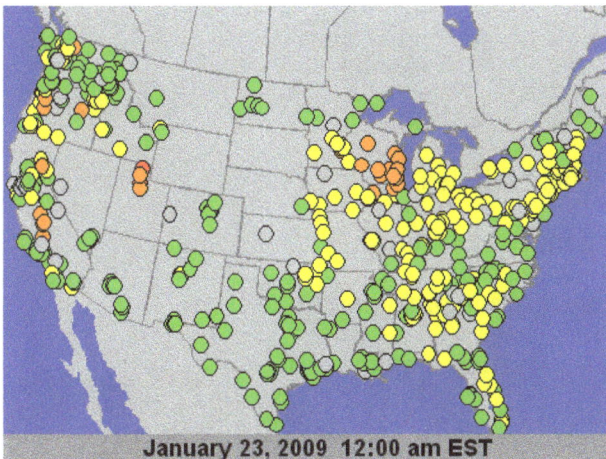

January 23, 2009 12:00 am EST

PM$_{2.5}$ 24-Hour AQI Loop, Courtesy US EPA

The United States Environmental Protection Agency (EPA) has developed an Air Quality Index that is used to report air quality. This AQI is divided into six categories indicating increasing levels of health concern. An AQI value over 300 represents hazardous air quality and below 50 the air quality is good.

Air Quality Index (AQI) Values	Levels of Health Concern	Colors
0 to 50	Good	Green
51 to 100	Moderate	Yellow
101 to 150	Unhealthy for Sensitive Groups	Orange
151 to 200	Unhealthy	Red
201 to 300	Very Unhealthy	Purple
301 to 500	Hazardous	Maroon

The AQI is based on the five "criteria" pollutants regulated under the Clean Air Act: ground-level ozone, particulate matter, carbon monoxide, sulfur dioxide, and nitrogen dioxide. The EPA has established National Ambient Air Quality Standards (NAAQS) for each of these pollutants in order to protect public health. An AQI value of 100 generally corresponds to the level of the NAAQS for the pollutant. The Clean Air Act (USA) (1990) requires EPA to review its National Ambient Air Quality Standards every five years to reflect evolving health effects information. The Air Quality Index is adjusted periodically to reflect these changes.

Computing the AQI

The air quality index is a piecewise linear function of the pollutant concentration. At the boundary between AQI categories, there is a discontinuous jump of one AQI unit. To convert from concentration to AQI this equation is used:

$$I = \frac{I_{high} - I_{low}}{C_{high} - C_{low}}(C - C_{low}) + I_{low}$$

where:

I = the (Air Quality) index,

C = the pollutant concentration,

C_{low} = the concentration breakpoint that is $\leq C$,

C_{high} = the concentration breakpoint that is $\geq C$,

I_{low} = the index breakpoint corresponding to C_{low},

I_{high} = the index breakpoint corresponding to C_{high}.

EPA's table of breakpoints is:

O_3 (ppb)	O_3 (ppb)	$PM_{2.5}$ ($\mu g/m^3$)	PM_{10} ($\mu g/m^3$)	CO (ppm)	SO_2 (ppb)	NO_2 (ppb)	AQI	AQI
C_{low} - C_{high} (avg)	C_{low} - C_{high} (avg)	C_{low} - C_{high} (avg)	C_{low} - C_{high} (avg)	C_{low} - C_{high} (avg)	C_{low} - C_{high} (avg)	C_{low} - C_{high} (avg)	I_{low} - I_{high}	Category
0-54 (8-hr)	-	0.0-12.0 (24-hr)	0-54 (24-hr)	0.0-4.4 (8-hr)	0-35 (1-hr)	0-53 (1-hr)	0-50	Good
55-70 (8-hr)	-	12.1-35.4 (24-hr)	55-154 (24-hr)	4.5-9.4 (8-hr)	36-75 (1-hr)	54-100 (1-hr)	51-100	Moderate
71-85 (8-hr)	125-164 (1-hr)	35.5-55.4 (24-hr)	155-254 (24-hr)	9.5-12.4 (8-hr)	76-185 (1-hr)	101-360 (1-hr)	101-150	Unhealthy for Sensitive Groups
86-105 (8-hr)	165-204 (1-hr)	55.5-150.4 (24-hr)	255-354 (24-hr)	12.5-15.4 (8-hr)	186-304 (1-hr)	361-649 (1-hr)	151-200	Unhealthy
106-200 (8-hr)	205-404 (1-hr)	150.5-250.4 (24-hr)	355-424 (24-hr)	15.5-30.4 (8-hr)	305-604 (24-hr)	650-1249 (1-hr)	201-300	Very Unhealthy
-	405-504 (1-hr)	250.5-350.4 (24-hr)	425-504 (24-hr)	30.5-40.4 (8-hr)	605-804 (24-hr)	1250-1649 (1-hr)	301-400	Hazardous
-	505-604 (1-hr)	350.5-500.4 (24-hr)	505-604 (24-hr)	40.5-50.4 (8-hr)	805-1004 (24-hr)	1650-2049 (1-hr)	401-500	

Suppose a monitor records a 24-hour average fine particle ($PM_{2.5}$) concentration of 12.0 micrograms per cubic meter. The equation above results in an AQI of:

$$\frac{50-0}{12.0-0}(12.0-0)+0 = 50,$$

corresponding to air quality in the "Good" range. To convert an air pollutant concentration to an AQI, EPA has developed a calculator.

If multiple pollutants are measured at a monitoring site, then the largest or "dominant" AQI value is reported for the location. The ozone AQI between 100 and 300 is computed by selecting the larger of the AQI calculated with a 1-hour ozone value and the AQI computed with the 8-hour ozone value.

8-hour ozone averages do not define AQI values greater than 300; AQI values of 301 or greater are calculated with 1-hour ozone concentrations. 1-hour SO_2 values do not define higher AQI values greater than 200. AQI values of 201 or greater are calculated with 24-hour SO_2 concentrations.

Real time monitoring data from continuous monitors are typically available as

1-hour averages. However, computation of the AQI for some pollutants requires averaging over multiple hours of data. (For example, calculation of the ozone AQI requires computation of an 8-hour average and computation of the $PM_{2.5}$ or PM_{10} AQI requires a 24-hour average.) To accurately reflect the current air quality, the multi-hour average used for the AQI computation should be centered on the current time, but as concentrations of future hours are unknown and are difficult to estimate accurately, EPA uses surrogate concentrations to estimate these multi-hour averages. For reporting the $PM_{2.5}$, PM_{10} and ozone air quality indices, this surrogate concentration is called the NowCast. The Nowcast is a particular type of weighted average that provides more weight to the most recent air quality data when air pollution levels are changing.

Public Availability of the AQI

Real time monitoring data and forecasts of air quality that are color-coded in terms of the air quality index are available from EPA's AirNow web site. Historical air monitoring data including AQI charts and maps are available at EPA's AirData website.

History of the AQI

The AQI made its debut in 1968, when the National Air Pollution Control Administration undertook an initiative to develop an air quality index and to apply the methodology to Metropolitan Statistical Areas. The impetus was to draw public attention to the issue of air pollution and indirectly push responsible local public officials to take action to control sources of pollution and enhance air quality within their jurisdictions.

Jack Fensterstock, the head of the National Inventory of Air Pollution Emissions and Control Branch, was tasked to lead the development of the methodology and to compile the air quality and emissions data necessary to test and calibrate resultant indices.

The initial iteration of the air quality index used standardized ambient pollutant concentrations to yield individual pollutant indices. These indices were then weighted and summed to form a single total air quality index. The overall methodology could use concentrations that are taken from ambient monitoring data or are predicted by means of a diffusion model. The concentrations were then converted into a standard statistical distribution with a preset mean and standard deviation. The resultant individual pollutant indices are assumed to be equally weighted, although values other than unity can be used. Likewise, the index can incorporate any number of pollutants although it was only used to combine SOx, CO, and TSP because of a lack of available data for other pollutants.

While the methodology was designed to be robust, the practical application for all metropolitan areas proved to be inconsistent due to the paucity of ambient air quality monitoring data, lack of agreement on weighting factors, and non-uniformity of air quality standards across geographical and political boundaries. Despite these issues, the publication of lists ranking metropolitan areas achieved the public policy objectives and led to the future development of improved indices and their routine application.

Air Quality Egg

The Air Quality Egg (AQE) is an Open source hardware Internet of Things platform and hobbyist device for crowdsourced citizen monitoring of airborne pollutants. The device won widespread recognition when it was named a Best of Kickstarter 2012 project, and has been featured in newspaper, magazine, peer-reviewed scientific journal and prominent blog articles. Crowdsourced data from citizen owners of the devices is uploaded to Xively where it becomes publicly accessibly on the Air Quality Egg website. The device is supported by 3rd party mobile apps such as Acculation's AQCalc.

The AQE grew out of Internet of Things meetup groups in New York City and Amsterdam, led by Pachube evangelist Ed Borden. WickedDevice LLC, the company that manufacturers and sells the AQE off its website, is a tech start-up headquartered in Ithaca, NY. There are two versions of the device: an Arduino shield for use by hobbyists, and a more consumer-ready "hobbyist kit" device. The latter consists of two identical-looking plastic enclosures vaguely resembling white eggs. One unit, the base unit, is connected to the user's ethernet LAN connection. The second unit monitors NO_2 and CO levels and reports these readings every few minutes back to the base unit via a custom wireless protocol. The base unit, in term, reports these readings to Xively and the AQE website. Add-ons are available for purchase off the website that add PM2.5 dust, Ozone, and VOC sensors.

Despite being labelled a not-consumer-ready "hobbyist" device by the manufacturer, the AQE is one of the few de facto commercially available, comprehensive Internet of Things pollution monitors on the US market. A number of competing devices have been announced, such as the Chemisense Wearable in the US or the Alima in France, but these do not appear to be commercially available in the US at time of writing. CitizenSensor is another US crowdsourced air quality monitoring device, but it is listed an "outgoing project", "DIY", and apparently not available for purchase. In terms of a comprehensive Internet of Things air pollution monitoring solution, the most direct competitor may be China's iKair (pronounced "I Care"), although this uses a closed, proprietary platform rather than the AQE's open source hardware and open data design. Unfortunately, at time of writing the iKair's companion smartphone app does not appear to be available in English or for download in the US. IBM recently announced a partnership with the government of China for analytics software to process data from pollution sensors, so both the software and hardware aspects of this market may be heating up. A major potential criticism of the AQE is that it is still a hobbyist device despite the appearance of consumer-ready packaging. In an effort to get these devices to end-users more quickly, the AQE's sensors have not yet been fully calibrated by the manufacturer. No doubt these issues will be addressed as more competing devices enter the market.

Ambient Air Quality Criteria

Ambient air quality criteria or standards are concentrations of pollutants in the air (usually outdoor air but sometimes indoor air) specified for a variety of reasons including for the protection of human health, buildings, crops, vegetation, ecosystems, as well as for planning and other purposes. There is no internationally accepted definition but usually "standards" have some legal or enforcement aspect, whereas "guidelines" may not be backed by laws. "Criteria/criterion" can be used as a generic term to cover standards and guidelines.

Various organisations have proposed criteria e.g. WHO, EU, US EPA and they are often similar - but not always, even if they are proposed for the same purpose (e.g. the protection of human health).

Specifying the Criteria

Important for any numerical standard is to ensure that averaging period, unit, and statistical measure are given (e.g. 98th percentile of hourlry means measured over a calendar year in micrograms per cubic metre ($\mu g/m^3$)). Without all of these three aspects the criterion is confusing and meaningless. Criteria can be set in different units (e.g. $\mu g/m^3$, parts per billion by volume (ppbv), parts per billion by mass (ppb(mass)), parts per million (ppm)) and it is possible to convert between all of these units if you know the molecular mass of the pollutant and the temperature at which you want to convert. Different temperatures are used throughout the world and so it is important to state the temperature of conversion (if relevant). Most pollutants have ambient criteria in the parts per billion (ppb)/$\mu g/m^3$ range. Some have smaller units (e.g. dioxins are often in pico grams /m^3); others have larger units (e.g. carbon monoxide (CO) in mg/m^3). Particle pollution (e.g. PM_{10}, $PM_{1.0}$) is specified in units of mass (e.g. $\mu g/m^3$) and not in units of volume (ppmv).

The Criteria

Below is a list of available criteria around the world. There is a lot of cross referencing (e.g. the International Finance Corp (IFC) has their own criteria but they are a copy of those specified by the WHO. It is important to check the reference as not all the related caveats/controlling parameters of the criterion can be put in the table. Also some criteria require certain specific ways of monitoring to demonstrate compliance.

Pollutant	Abbr.	Value	Units	Averaging Period	Statistical Measure	# exceedences	Organisation	Status	Applicability	Notes
Nitrogen dioxide	NO_2	40	$\mu g/m^3$	Annual	Mean	None	WHO	Guideline	Global	
Nitrogen dioxide	NO_2	40	$\mu g/m^3$	Annual	Mean	None	EU	Limit Value	EU Member States	Calendar Year
Nitrogen dioxide	NO_2	100	$\mu g/m^3$	Annual	Mean	None	US EPA	Legal	United States	Quoted as 53 ppb
Nitrogen dioxide	NO_2	113	$\mu g/m^3$	Daily	Daily mean of hourly	None	Japanese Ministry of Environment	Legal	Japan	Quoted as 0.06 ppm
Nitrogen dioxide	NO_2	100	$\mu g/m^3$	Annual	Yrly mean of daily	None	SCENR	Legal	Qatar	Calendar Year
Nitrogen dioxide	NO_2	150	$\mu g/m^3$	Daily (24hr)	99.7%ile	1 day/yr	SCENR	Legal	Qatar	Calendar Year
Nitrogen dioxide	NO_2	200	$\mu g/m^3$	1 hour	99.79%ile	18 hrs/yr	EU	Limit value	EU Member States	Calendar Year
Nitrogen dioxide	NO_2	400	$\mu g/m^3$	1 hour	99.7%ile	26 hrs/yr	SCENR	Legal	Qatar	Calendar Year

Pollutant	Abbr.	Value	Units	Averaging Period	Statistical Measure	# exceedences	Organisation	Status	Applicability	Notes
Sulphur dioxide	SO_2	125	$\mu g/m^3$	24 hour	Any 24 hour period	None	WHO	Interim Target (IT) 1	Global	Was the guideline in 2000
Sulphur dioxide	SO_2	50	$\mu g/m^3$	24 hour	Any 24 hour period	None	WHO	Interim Target (IT) 2	Global	Calendar Year (1 Jan – 31 Dec)
Sulphur dioxide	SO_2	125	$\mu g/m^3$	24 hour	Any 24 hour mean	None	WHO	Interim Target (IT) 1	Global	

Pollutant	Abbr.	Value	Units	Averaging Period	Statistical Measure	# exceedences	Organisation	Status	Applicability	Notes
Sulphur dioxide	SO_2	50	µg/m³	24 hour	Any 24 hour mean	None	WHO	Interim Target (IT) 2	Global	
Sulphur dioxide	SO_2	20	µg/m³	24 hour	Any 24 hour mean	None	WHO	Guideline	Global	
Sulphur dioxide	SO_2	500	µg/m³	10 minute	Any 10 minute mean	None	WHO	Guideline	Global	
Sulphur dioxide	SO_2	1300	µg/m³	1 hour	99.99%ile	1 hr/year	Ras Laffan	Legal	Qatar	Calendar Year

Pollutant	Abbr.	Value	Units	Averaging Period	Statistical Measure	# exceedences	Organisation	Status	Applicability	Notes
Fine particles	PM_{10}	125	µg/m³	24 hour	99th %ile	3 days/yr	WHO	Guideline	Global	
Very fine particles	$PM_{2.5}$	25	µg/m³	24 hour	Any 24 hr mean	None	WHO	Guideline	Global	
Very fine particles	$PM_{2.5}$	75	µg/m³	24 hour	Any 24 hr mean	None	WHO	Interim target (IT) 1	Global	
Very fine particles	$PM_{2.5}$	50	µg/m³	24 hour	Any 24 hr mean	None	WHO	Interim target (IT) 2	Global	
Very fine particles	$PM_{2.5}$	37.5	µg/m³	24 hour	Any 24 hr mean	None	WHO	Interim target (IT) 3	Global	
Very fine particles	$PM_{2.5}$	10	µg/m³	Annual	Mean	None	WHO	Guideline	Global	
Very fine particles	$PM_{2.5}$	35	µg/m³	Annual	Mean	None	WHO	Interim target (IT) 1	Global	

	Abbr.	Value	Units	Averaging Period	Statistical Measure	# exceedences	Organisation	Status	Applicability	Notes
Very fine particles	PM$_{2.5}$	25	µg/m³	Annual	Mean	None	WHO	Interim target (IT) 2	Global	
Very fine particles	PM$_{2.5}$	15	µg/m³	Annual	Mean	None	WHO	Interim target (IT) 3	Global	

Pollutant	Abbr.	Value	Units	Averaging Period	Statistical Measure	# exceedences	Organisation	Status	Applicability	Notes
Ozone	O$_3$	100	µg/m³	8 hours	Any 8 hour mean	None	WHO	Guideline	Global	
Ozone	O$_3$	160	µg/m³	8 hours	Any 8 hour mean	None	WHO	Interim target	Global	
Ozone	O$_3$	235	µg/m³	1 hour	99.9%	8 hrs/yr	SCENR	Legal	Qatar	Calendar Year
Ozone	O$_3$	120	µg/m³	8 hours	99.8%	2x8hrs/yr	SCENR	Legal	Qatar	Calendar Year

Pollutant	Abbr.	Value	Units	Averaging Period	Statistical Measure	# exceedences	Organisation	Status	Applicability	Notes
Carbon monoxide	CO	40000	µg/m³	1 hour	Maximum	None	SCENR	Legal	Qatar	Calendar Year
Carbon monoxide	CO	10000	µg/m³	8 hour	Maximum	None	SCENR	Legal	Qatar	Calendar Year

Continuous Emissions Monitoring System

Continuous emission monitoring systems (CEMS) were historically used as a tool to monitor flue gas for oxygen, carbon monoxide and carbon dioxide to provide information for combustion control in industrial settings. They are currently used as a means to comply with air emission standards such as the United States Environmental Protection Agency's Acid Rain Program, other federal emission programs, or state permitted emission standards. Facilities employ the use of CEMS to continuously collect, record and report the required emissions data.

The standard CEM system consists of a sample probe, filter, sample line (umbilical), gas conditioning system, calibration gas system, and a series of gas analyzers which reflect the parameters being monitored. Typical monitored emissions include: sulfur dioxide, nitrogen oxides, carbon monoxide, carbon dioxide, hydrogen chloride, airborne particulate matter, mercury, volatile organic compounds, and oxygen. CEM systems can also measure air flow, flue gas opacity and moisture.

In the U.S., the EPA requires a data acquisition and handling system to collect and report the data. SO2 emissions must be measured in pounds per hour using both an SO2 pollutant concentration monitor and a volumetric flow monitor. For NO_x, both a NO_x pollutant concentration monitor and a diluent gas monitor are required to determine the emissions rate (lbs/mmBtu). Opacity must also be monitored. NO_x measuring is not a current requirement, however if monitored, a CO_2 or oxygen monitor plus a flow monitor should be used. In monitoring these emissions, the system must be in continuous operation and must be able to sample, analyze, and record data at least every 15 minutes and then averaged hourly.

Operation

A small sample of flue gas is extracted, by means of a pump, into the CEM system via a sample probe. Facilities that combust fossil fuels often use a dilution-extractive probe to dilute the sample with clean, dry air to a ratio typically between 50:1 to 200:1, but usually 100:1. Dilution is used because pure flue gas can be hot, wet and, with some pollutants, sticky. Once diluted to the appropriate ratio, the sample is transported through a sample line (typically referred to as an umbilical) to a manifold from which individual analyzers may extract a sample. Gas analyzers employ various techniques to accurately measure concentrations. Some commonly used techniques include: infrared and ultraviolet adsorption, chemiluminescence, fluorescence and beta ray absorption. After analysis, the gas exits the analyzer to a common manifold to all analyzers where it is vented out of doors. A Data Acquisition and Handling System (DAHS) receives the signal output from each analyzer in order to collect and record emissions data.

Another sample extraction method used in industrial sources and utility sources with low emission rates, is commonly referred to as the "hot dry" extractive method or "direct" CEMS. The sample is not diluted, but is carried along a heated sample line at high temperature into a sample conditioning unit. The sample is filtered to remove particulate matter and dried, usually with a chiller, to remove moisture. Once conditioned, the sample enters a sampling manifold and is measured using the same methods above. One advantage of this method is the ability to measure % oxygen in the sample, which is often required in the regulatory calculations for emission corrections. Since dilution mixes clean dry air with the sample, dilution systems cannot measure % oxygen.

Quality Assurance

Accuracy of the system is demonstrated in several ways. An internal quality assurance check is achieved by daily introduction of a certified concentration of gas to the sample probe. The EPA also allows for the use of Continuous Emissions Monitoring Calibration Systems which dilute gases to generate calibration standards. The analyzer reading must be accurate to a certain percentage. The percent accuracy can vary, but most fall between 2.5% and 5%. In power stations affected by the Acid Rain Program, annual (or bi-annual) certification of the system must be performed by an independent firm. The firm will have an independent CEM system temporarily in place to collect emissions data in parallel with the plant CEMS. This testing is referred to as a Relative Accuracy Test Audit (RATA).

In the U.S., periodic evaluations of the equipment must be reported and recorded. This includes daily calibration error tests, daily interference tests for flow monitors, and semi-annual (or annual) RATA and bias tests. CEMS equipment is expensive and not always affordable for a facility. In such cases, a facility will install non-EPA compliant analysis equipment at the emissions point. Once yearly, for the equipment evaluation, a mobile CEMS company measures emissions with compliant equipment. The results are then compared to the non-compliant analyzer system.

References

- Davis, Devra (2002). When Smoke Ran Like Water: Tales of Environmental Deception and the Battle Against Pollution. Basic Books. ISBN 0-465-01521-2.

- Beychok, M.R. (2005). Fundamentals of Stack Gas Dispersion (4th ed.). author-published. ISBN 0-9644588-0-2. www.air-dispersion.com

- Turner, D.B. (1994). Workbook of atmospheric dispersion estimates: an introduction to dispersion modeling (2nd ed.). CRC Press. ISBN 1-56670-023-X.

- Garcia, Javier; Colosio, Joëlle (2002). Air-quality indices : elaboration, uses and international comparisons. Presses des MINES. ISBN 2-911762-36-3.

- The Babcock & Wilcox Company. Steam: its generation and use. The Babcock & Wilcox Company. pp. 36–5. ISBN 0-9634570-1-2.

- "Study Links 6.5 Million Deaths Each Year to Air Pollution". New York Times. 26 June 2016. Retrieved 27 June 2016.

- "Specifications and Test Procedures for Total Hydrocarbon Continuous Monitoring Systems in Stationary Sources" (PDF). www3.epa.gov. Retrieved 23 February 2016.

- "Bucknell tent death: Hannah Thomas-Jones died from carbon monoxide poisoning". BBC News. 17 January 2013. Retrieved 22 September 2015.

- St. Fleur, Nicholas (10 November 2015). "Atmospheric Greenhouse Gas Levels Hit Record, Report Says". New York Times. Retrieved 11 November 2015.

- Ritter, Karl (9 November 2015). "UK: In 1st, global temps average could be 1 degree C higher". AP News. Retrieved 11 November 2015.

- Cole, Steve; Gray, Ellen (14 December 2015). "New NASA Satellite Maps Show Human Fingerprint on Global Air Quality". NASA. Retrieved 14 December 2015.

Air Pollution Dispersion: A Comprehensive Study

Air pollution dispersion is the amount of pollution that is distributed into the atmosphere. Air pollution comes from anthropogenic or natural sources. The aspects elucidated are atmospheric dispersion modeling, roadway air dispersion modeling, ADMS 3, CALPUFF and NAME. The section is an overview of the major aspects of air pollution dispersion.

Outline of Air Pollution Dispersion

Air pollution dispersion – distribution of air pollution into the atmosphere. Air pollution is the introduction of particulates, biological molecules, or other harmful materials into Earth's atmosphere, causing disease, death to humans, damage to other living organisms such as food crops, or the natural or built environment. Air pollution may come from anthropogenic or natural sources. Dispersion refers to what happens to the pollution during and after its introduction; understanding this may help in identifying and controlling it. Air pollution dispersion has become the focus of environmental conservationists and governmental environmental protection agencies (local, state, province and national) of many countries (which have adopted and used much of the terminology of this field in their laws and regulations) regarding air pollution control.

Air Pollution Emission Plumes

Air pollution emission plume – flow of pollutant in the form of vapor or smoke released into the air. Plumes are of considerable importance in the atmospheric dispersion modelling of air pollution. There are three primary types of air pollution emission plumes:

- Buoyant plumes — Plumes which are lighter than air because they are at a higher temperature and lower density than the ambient air which surrounds them, or because they are at about the same temperature as the ambient air but have a lower molecular weight and hence lower density than the ambient air. For example, the emissions from the flue gas stacks of industrial furnaces are buoyant because they are considerably warmer and less dense than the ambient air. As another example, an emission plume of methane gas at ambient air temperatures is buoyant because methane has a lower molecular weight than the ambient air.

- Dense gas plumes — Plumes which are heavier than air because they have a higher density than the surrounding ambient air. A plume may have a higher density than air because it has a higher molecular weight than air (for example, a plume of carbon dioxide). A plume may also have a higher density than air if the plume is at a much lower temperature than the air. For example, a plume of evaporated gaseous methane from an accidental release of liquefied natural gas (LNG) may be as cold as -161 °C.

- Passive or neutral plumes — Plumes which are neither lighter or heavier than air.

Air Pollution Dispersion Models

There are five types of air pollution dispersion models, as well as some hybrids of the five types:

- Box model — The box model is the simplest of the model types. It assumes the airshed (i.e., a given volume of atmospheric air in a geographical region) is in the shape of a box. It also assumes that the air pollutants inside the box are homogeneously distributed and uses that assumption to estimate the average pollutant concentrations anywhere within the airshed. Although useful, this model is very limited in its ability to accurately predict dispersion of air pollutants over an airshed because the assumption of homogeneous pollutant distribution is much too simple.

- Gaussian model — The Gaussian model is perhaps the oldest (circa 1936) and perhaps the most commonly used model type. It assumes that the air pollutant dispersion has a Gaussian distribution, meaning that the pollutant distribution has a normal probability distribution. Gaussian models are most often used for predicting the dispersion of continuous, buoyant air pollution plumes originating from ground-level or elevated sources. Gaussian models may also be used for predicting the dispersion of non-continuous air pollution plumes (called *puff models*). The primary algorithm used in Gaussian modeling is the *Generalized Dispersion Equation For A Continuous Point-Source Plume*.

- Lagrangian model — a Lagrangian dispersion model mathematically follows pollution plume parcels (also called particles) as the parcels move in the atmosphere and they model the motion of the parcels as a random walk process. The Lagrangian model then calculates the air pollution dispersion by computing the statistics of the trajectories of a large number of the pollution plume parcels. A Lagrangian model uses a moving frame of reference as the parcels move from their initial location. It is said that an observer of a Lagrangian model follows along with the plume.

- Eulerian model — an Eulerian dispersions model is similar to a Lagrangian model in that it also tracks the movement of a large number of pollution plume parcels as they move from their initial location. The most important difference between the two models is that the Eulerian model uses a fixed three-dimensional Cartesian grid as a frame of reference rather than a moving frame of reference. It is said that an observer of an Eulerian model watches the plume go by.

- Dense gas model — Dense gas models are models that simulate the dispersion of dense gas pollution plumes (i.e., pollution plumes that are heavier than air). The three most commonly used dense gas models are:

    o The DEGADIS model developed by Dr. Jerry Havens and Dr. Tom Spicer at the University of Arkansas under commission by the US Coast Guard and US EPA.

    o The SLAB model developed by the Lawrence Livermore National Laboratory funded by the US Department of Energy, the US Air Force and the American Petroleum Institute.

    o The HEGADAS model developed by Shell Oil's research division.

Air Pollutant Emission

Air pollution emission source

- Types of air pollutant emission sources – named for their characteristics

    o Sources, by shape – there are four basic shapes which an emission source may have. They are:

        o Point source — single, identifiable source of air pollutant emissions (for example, the emissions from a combustion furnace flue gas stack). Point sources are also characterized as being either elevated or at ground-level. A point source has no geometric dimensions.

- o Line source — one-dimensional source of air pollutant emissions (for example, the emissions from the vehicular traffic on a roadway).

- o Area source — two-dimensional source of diffuse air pollutant emissions (for example, the emissions from a forest fire, a landfill or the evaporated vapors from a large spill of volatile liquid).

- o Volume source — three-dimensional source of diffuse air pollutant emissions. Essentially, it is an area source with a third (height) dimension (for example, the fugitive gaseous emissions from piping flanges, valves and other equipment at various heights within industrial facilities such as oil refineries and petrochemical plants). Another example would be the emissions from an automobile paint shop with multiple roof vents or multiple open windows.

o Sources, by motion

- o Stationary source – flue gas stacks are examples of stationary sources

- o Mobile source – buses are examples of mobile sources

o Sources, by urbanization level – whether the source is within a city or not is relevant in that urban areas constitute a so-called *heat island* and the heat rising from an urban area causes the atmosphere above an urban area to be more turbulent than the atmosphere above a rural area

- o Urban source – emission is in an urban area

- o Rural source – emission is in a rural area

o Sources, by elevation

- o Surface or ground-level source

- o Near surface source

- o Elevated source

o Sources, by duration

- o Puff or intermittent source – short term sources (for example, many accidental emission releases are short term puffs)

- o Continuous source – long term source (for example, most flue gas stack emissions are continuous)

Characterization of Atmospheric Turbulence

Effect of turbulence on dispersion – turbulence increases the entrainment and mixing of unpolluted air into the plume and thereby acts to reduce the concentration of pollutants in the plume (i.e., enhances the plume dispersion). It is therefore important to categorize the amount of atmospheric turbulence present at any given time...

The Pasquill Atmospheric Stability Classes

Pasquill atmospheric stability classes – oldest and, for a great many years, the most commonly used method of categorizing the amount of atmospheric turbulence present was the method developed by Pasquill in 1961. He categorized the atmospheric turbulence into six stability classes named A, B, C, D, E and F with class A being the most unstable or most turbulent class, and class F the most stable or least turbulent class. Table 1 lists the six classes and Table 2 provides the meteorological conditions that define each class.

Table 1: The Pasquill Stability Classes

Stability class	Definition		Stability class	Definition
A	very unstable		D	neutral
B	unstable		E	slightly stable
C	slightly unstable		F	stable

Table 2: Meteorological Conditions that Define the Pasquill Stability Classes

Surface windspeed		Daytime incoming solar radiation			Nighttime cloud cover	
m/s	mi/h	Strong	Moderate	Slight	> 50%	< 50%
< 2	< 5	A	A – B	B	E	F
2 – 3	5 – 7	A – B	B	C	E	F
3 – 5	7 – 11	B	B – C	C	D	E
5 – 6	11 – 13	C	C – D	D	D	D
> 6	> 13	C	D	D	D	D
Note: Class D applies to heavily overcast skies, at any windspeed day or night						

Data Availability

- Historical stability class data – known as the Stability Array (STAR) data, for sites within the USA can be purchased from the National Climatic Data Center (NCDC).

Advanced Methods of Categorizing Atmospheric Turbulence

Advanced air pollution dispersion models – they do not categorize atmospheric turbulence by using the simple meteorological parameters commonly used in defining the six Pasquill classes as shown in Table 2 above. The more advanced models use some form of Monin-Obukhov similarity theory. Some examples include:

- AERMOD – US EPA's most advanced model, no longer uses the Pasquill stability classes to categorize atmospheric turbulence. Instead, it uses the surface roughness length and the Monin-Obukhov length.

- ADMS 4, – United Kingdom's most advanced model, uses the Monin-Obukhov length, the boundary layer height and the windspeed to categorize the atmospheric turbulence.

Miscellaneous other Terminology

- Building effects or downwash: When an air pollution plume flows over nearby buildings or other structures, turbulent eddies are formed in the downwind side of the building. Those eddies cause a plume from a stack source located within about five times the height of a nearby building or structure to be forced down to the ground much sooner than it would if a building or structure were not present. The effect can greatly increase the resulting near-by ground-level pollutant concentrations downstream of the building or structure. If the pollutants in the plume are subject to depletion by contact with the ground (particulates, for example), the concentration increase just downstream of the building or structure will decrease the concentrations further downstream.

- Deposition of the pollution plume components to the underlying surface can be defined as either dry or wet deposition:

  o   Dry deposition is the removal of gaseous or particulate material from the pollution plume by contact with the ground surface or vegetation (or even water surfaces) through transfer processes such as absorption and gravitational sedimentation. This may be calculated by means of a *deposition velocity*, which is related to the resistance of the underlying surface to the transfer.

  o   Wet deposition is the removal of pollution plume components by the action of rain. The wet deposition of radionuclides in a pollution plume by a burst of rain often forms so called *hot spots* of radioactivity on the underlying surface.

- Inversion layers: Normally, the air near the Earth's surface is warmer than the air above it because the atmosphere is heated from below as solar radiation warms the Earth's surface, which in turn then warms the layer of the atmosphere directly above it. Thus, the atmospheric temperature normally decreases with increasing altitude. However, under certain meteorological conditions, atmospheric layers may form in which the temperature increases with increasing altitude. Such layers are called inversion layers. When such a layer forms at the Earth's surface, it is called a surface inversion. When an inversion layer forms at some distance above the earth, it is called an inversion aloft (sometimes referred to as a *capping inversion*). The air within an inversion aloft is very stable with very little vertical motion. Any rising parcel of air within the inversion soon expands, thereby adiabatically cooling to a lower temperature than the surrounding air and the parcel stops rising. Any sinking parcel soon compresses adiabatically to a higher temperature than the surrounding air and the parcel stops sinking. Thus, any air pollution plume that enters an inversion aloft will undergo very little vertical mixing unless it has sufficient momentum to completely pass through the inversion aloft. That is one reason why an inversion aloft is sometimes called a capping inversion.

- Mixing height: When an inversion aloft is formed, the atmospheric layer between the Earth's surface and the bottom of the inversion aloft is known as the mixing layer and the distance between the Earth's surface and the bottom of inversion aloft is known as the mixing height. Any air pollution plume dispersing beneath an inversion aloft will be limited in vertical mixing to that which occurs beneath the bottom of the inversion aloft (sometimes called the *lid*). Even if the pollution plume penetrates the inversion, it will not undergo any further significant vertical mixing. As for a pollution plume passing completely through an inversion layer aloft, that rarely occurs unless the pollution plume's source stack is very tall and the inversion lid is fairly low.

Atmospheric Dispersion Modeling

Atmospheric dispersion modeling is the mathematical simulation of how air pollutants disperse in the ambient atmosphere. It is performed with computer programs that solve the mathematical equations and algorithms which simulate the pollutant dispersion. The dispersion models are used to estimate the downwind ambient concentration of air pollutants or toxins emitted from sources such as industrial plants, vehicular traffic or accidental chemical releases. They can also be used to predict future concentrations under specific scenarios (i.e. changes in emission sources). Therefore, they are the dominant type of model used in air quality policy making. They are most useful for pollutants that are dispersed over large distances and that may react in the atmosphere. For pollutants that have a very high spatio-temporal variability (i.e. have very steep distance to source decay such as black carbon) and for epidemiological studies statistical land-use regression models are also used.

Dispersion models are important to governmental agencies tasked with protecting and managing the ambient air quality. The models are typically employed to determine whether existing or proposed new industrial facilities are or will be in compliance with the National Ambient Air Quality Standards (NAAQS) in the United States and other nations. The models also serve to assist in the design of effective control strategies to reduce emissions of harmful air pollutants. During the late 1960s, the Air Pollution Control Office of the U.S. EPA initiated research projects that would lead to the development of models for the use by urban and transportation planners. A major and significant application of a roadway dispersion model that resulted from such research was applied to the Spadina Expressway of Canada in 1971.

Air dispersion models are also used by public safety responders and emergency management personnel for emergency planning of accidental chemical releases. Models are used to determine the consequences of accidental releases of hazardous or toxic materials, Accidental releases may result in fires, spills or explosions that involve hazardous materials, such as chemicals or radionuclides. The results of dispersion modeling, using worst case accidental release source terms and meteorological conditions, can provide an estimate of location impacted areas, ambient concentrations, and be used to determine protective actions appropriate in the event a release occurs. Appropriate protective actions may include evacuation or shelter in place for persons in the downwind direction. At industrial facilities, this type of consequence assessment or emergency planning is required under the Clean Air Act (United States) (CAA) codified in Part 68 of Title 40 of the Code of Federal Regulations.

The dispersion models vary depending on the mathematics used to develop the model, but all require the input of data that may include:

- Meteorological conditions such as wind speed and direction, the amount of atmospheric turbulence (as characterized by what is called the "stability class"), the ambient air temperature, the height to the bottom of any inversion aloft that may be present, cloud cover and solar radiation.

- Source term (the concentration or quantity of toxins in emission or accidental release source terms) and temperature of the material

- Emissions or release parameters such as source location and height, type of source (i.e., fire, pool or vent stack)and exit velocity, exit temperature and mass flow rate or release rate.

- Terrain elevations at the source location and at the receptor location(s), such as nearby homes, schools, businesses and hospitals.

- The location, height and width of any obstructions (such as buildings or other structures) in the path of the emitted gaseous plume, surface roughness or the use of a more generic parameter "rural" or "city" terrain.

Many of the modern, advanced dispersion modeling programs include a pre-processor module for the input of meteorological and other data, and many also include a post-processor module for graphing the output data and/or plotting the area impacted by the air pollutants on maps. The plots of areas impacted may also include isopleths showing areas of minimal to high concentrations that define areas of the highest health risk. The isopleths plots are useful in determining protective actions for the public and responders.

The atmospheric dispersion models are also known as atmospheric diffusion models, air dispersion models, air quality models, and air pollution dispersion models.

Atmospheric Layers

Discussion of the layers in the Earth's atmosphere is needed to understand where airborne pollutants disperse in the atmosphere. The layer closest to the Earth's surface is known as the *troposphere*. It extends from sea-level to a height of about 18 km and contains about 80 percent of the mass of the overall atmosphere. The *stratosphere* is the next layer and extends from 18 km to about 50 km. The third layer is the *mesosphere* which extends from 50 km to about 80 km. There are other layers above 80 km, but they are insignificant with respect to atmospheric dispersion modeling.

The lowest part of the troposphere is called the *atmospheric boundary layer (ABL)* or the *planetary boundary layer (PBL)* and extends from the Earth's surface to about 1.5 to 2.0 km in height. The air temperature of the atmospheric boundary layer decreases with increasing altitude until it reaches what is called the *inversion layer* (where the temperature increases with increasing altitude) that caps the atmospheric boundary layer. The upper part of the troposphere (i.e., above the inversion layer) is called the *free troposphere* and it extends up to the 18 km height of the troposphere.

The ABL is of the most important with respect to the emission, transport and dispersion of airborne pollutants. The part of the ABL between the Earth's surface and the bottom of the inversion layer is known as the mixing layer. Almost all of the airborne pollutants emitted into the ambient atmosphere are transported and dispersed within the mixing layer. Some of the emissions penetrate the inversion layer and enter the free troposphere above the ABL.

In summary, the layers of the Earth's atmosphere from the surface of the ground upwards are: the ABL made up of the mixing layer capped by the inversion layer; the free troposphere; the stratosphere; the mesosphere and others. Many atmospheric dispersion models are referred to as *boundary layer models* because they mainly model air pollutant dispersion within the ABL. To avoid confusion, models referred to as *mesoscale models* have dispersion modeling capabilities that extend horizontally up to a few hundred kilometres. It does not mean that they model dispersion in the mesosphere.

Gaussian air Pollutant Dispersion Equation

The technical literature on air pollution dispersion is quite extensive and dates back to the 1930s and earlier. One of the early air pollutant plume dispersion equations was derived by Bosanquet and Pearson. Their equation did not assume Gaussian distribution nor did it include the effect of ground reflection of the pollutant plume.

Sir Graham Sutton derived an air pollutant plume dispersion equation in 1947 which did include the assumption of Gaussian distribution for the vertical and crosswind dispersion of the plume and also included the effect of ground reflection of the plume.

Under the stimulus provided by the advent of stringent environmental control regulations, there was an immense growth in the use of air pollutant plume dispersion calculations between the late 1960s and today. A great many computer programs for calculating the dispersion of air pollutant emissions were developed during that period of time and they were called "air dispersion models". The basis for most of those models was the Complete Equation For Gaussian Dispersion Modeling Of Continuous, Buoyant Air Pollution Plumes shown below:

$$C = \frac{Q}{u} \cdot \frac{f}{\sigma_y \sqrt{2\pi}} \cdot \frac{g_1 + g_2 + g_3}{\sigma_z \sqrt{2\pi}}$$

where:

f = crosswind dispersion parameter

$= \exp[-y^2 / (2\sigma_y^2)]$

g = vertical dispersion parameter = $g_1 + g_2 + g_3$

g_1 = vertical dispersion with no reflections

$= \exp[-(z-H)^2 / (2\sigma_z^2)]$

g_2 = vertical dispersion for reflection from the ground

$= \exp[-(z+H)^2 / (2\sigma_z^2)]$

g_3 = vertical dispersion for reflection from an inversion aloft

$= \sum_{m=1}^{\infty} \{ \exp[-(z-H-2mL)^2 / (2\sigma_z^2)]$

? $+ \exp[-(z+H+2mL)^2 / (2\sigma_z^2)]$

? $+ \exp[-(z+H-2mL)^2 / (2\sigma_z^2)]$

? $+ \exp[-(z-H+2mL)^2 / (2\sigma_z^2)]\}$

C = concentration of emissions, in g/m³, at any receptor located:

       x meters downwind from the emission source point

       y meters crosswind from the emission plume centerline

       z meters above ground level

Q = source pollutant emission rate, in g/s

u = horizontal wind velocity along the plume centerline, m/s

H = height of emission plume centerline above ground level, in m

σ_z = vertical standard deviation of the emission distribution, in m

σ_y = horizontal standard deviation of the emission distribution, in m

L = height from ground level to bottom of the inversion aloft, in m

exp = the exponential function

The above equation not only includes upward reflection from the ground, it also includes downward reflection from the bottom of any inversion lid present in the atmosphere.

The sum of the four exponential terms in g_3 converges to a final value quite rapidly. For most cases, the summation of the series with $m = 1$, $m = 2$ and $m = 3$ will provide an adequate solution.

σ_z and σ_y are functions of the atmospheric stability class (i.e., a measure of the turbulence in the ambient atmosphere) and of the downwind distance to the receptor. The two most important variables affecting the degree of pollutant emission dispersion obtained are the height of the emission source point and the degree of atmospheric turbulence. The more turbulence, the better the degree of dispersion.

The resulting calculations for air pollutant concentrations are often expressed as an air pollutant concentration contour map in order to show the spatial variation in contaminant levels over a wide area under study. In this way the contour lines can overlay sensitive receptor locations and reveal the spatial relationship of air pollutants to areas of interest.

Whereas older models rely on stability classes for the determination of σ_y and σ_z, more recent models increasingly rely on the Monin-Obukhov similarity theory to derive these parameters.

Briggs Plume Rise Equations

The Gaussian air pollutant dispersion equation requires the input of H which is the pollutant plume's centerline height above ground level—and H is the sum of H_s (the actual physical height of the pollutant plume's emission source point) plus ΔH (the plume rise due the plume's buoyancy).

To determine ΔH, many if not most of the air dispersion models developed between the late 1960s and the early 2000s used what are known as "the Briggs equations." G.A. Briggs first published his plume rise observations and comparisons in 1965. In 1968, at a symposium sponsored by CONCAWE (a Dutch organization), he compared many of the plume rise models then available in the literature. In that same year, Briggs also wrote the section of the publication edited by Slade dealing with the comparative analyses of plume rise models. That was followed in 1969 by his classical critical review of the entire plume rise literature, in which he proposed a set of plume rise equations which have become widely known as "the Briggs equations". Subsequently, Briggs modified his 1969 plume rise equations in 1971 and in 1972.

Briggs divided air pollution plumes into these four general categories:

- Cold jet plumes in calm ambient air conditions

- Cold jet plumes in windy ambient air conditions

- Hot, buoyant plumes in calm ambient air conditions

- Hot, buoyant plumes in windy ambient air conditions

Briggs considered the trajectory of cold jet plumes to be dominated by their initial velocity momentum, and the trajectory of hot, buoyant plumes to be dominated by their buoyant momentum to the extent that their initial velocity momentum was relatively unimportant. Although Briggs proposed plume rise equations for each of the above plume categories, *it is important to emphasize that "the Briggs equations" which become widely used are those that he proposed for bent-over, hot buoyant plumes.*

In general, Briggs's equations for bent-over, hot buoyant plumes are based on observations and data involving plumes from typical combustion sources such as the flue gas stacks from steam-generating boilers burning fossil fuels in large power plants. Therefore, the stack exit velocities were probably in the range of 20 to 100 ft/s (6 to 30 m/s) with exit temperatures ranging from 250 to 500 °F (120 to 260 °C).

A logic diagram for using the Briggs equations to obtain the plume rise trajectory of bent-over buoyant plumes is presented below:

where:	
Δh	= plume rise, in m
F	= buoyancy factor, in $m^4 s^{-3}$
x	= downwind distance from plume source, in m
x_f	= downwind distance from plume source to point of maximum plume rise, in m

| **u** | = windspeed at actual stack height, in m/s |
| **s** | = stability parameter, in s^{-2} |

LOGIC DIAGRAM FOR BRIGGS' EQUATIONS TO CALCULATE
THE RISE OF A BUOYANT PLUME

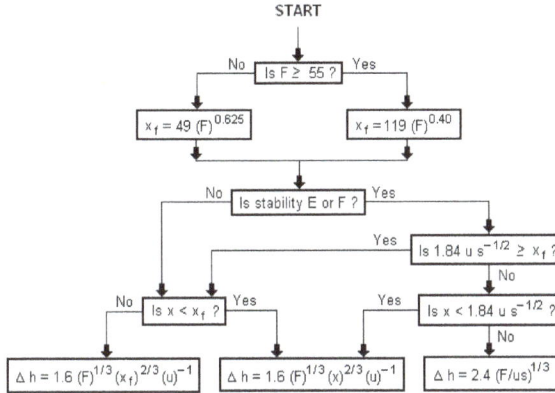

START

No ─── Is F ≥ 55 ? ─── Yes

$x_f = 49 \, (F)^{0.625}$ $x_f = 119 \, (F)^{0.40}$

No ─── Is stability E or F ? ─── Yes

Yes ─── Is $1.84 \, u \, s^{-1/2} \geq x_f$?

No

No ─── Is $x < x_f$? ─── Yes   Yes ─── Is $x < 1.84 \, u \, s^{-1/2}$?

No

$\Delta h = 1.6 \, (F)^{1/3} \, (x_f)^{2/3} \, (u)^{-1}$ $\Delta h = 1.6 \, (F)^{1/3} \, (x)^{2/3} \, (u)^{-1}$ $\Delta h = 2.4 \, (F/us)^{1/3}$

The above parameters used in the Briggs' equations are discussed in Beychok's book.

Roadway Air Dispersion Modeling

Roadway air dispersion is applied to highway segments

Roadway air dispersion modeling is the study of air pollutant transport from a roadway or other linear emitter. Computer models are required to conduct this analysis, because of the complex variables involved, including vehicle emissions, vehicle speed, meteorology, and terrain geometry. Line source dispersion has been studied since at least the 1960s, when the regulatory framework in the United States began requiring quantitative analysis of the air pollution consequences of major roadway and airport projects. By the early 1970s this subset of atmospheric dispersion models were being applied to real world cases of highway planning, even including some controversial court cases.

How the Model Works

The basic concept of the roadway air dispersion model is to calculate air pollutant levels in the vicinity of a highway or arterial roadway by considering them as line sources. The model takes into account source characteristics such as traffic volume, vehicle speeds, truck mix, and fleet emission controls; in addition, the roadway geometry, surrounding terrain and local meteorology are addressed. For example, many air quality standards require that certain near worst case meteorological conditions be applied.

The calculations are sufficiently complex that a computer model is essential to arrive at authoritative results, although workbook type manuals have been developed as screening techniques. In some cases where results must be refereed (such as legal cases), model validation may be needed with field test data in the local setting; this step is not usually warranted, because the best models have been extensively validated over a wide spectrum of input data variables.

The product of the calculations is usually a set of isopleths or mapped contour lines either in plan view or cross sectional view. Typically these might be stated as concentrations of carbon monoxide, total reactive hydrocarbons, oxides of nitrogen, particulate or benzene. The air quality scientist can run the model successively to study techniques of reducing adverse air pollutant concentrations (for example, by redesigning roadway geometry, altering speed controls or limiting certain types of trucks). The model is frequently utilized in an Environmental Impact Statement involving a major new roadway or land use change which will induce new vehicular traffic.

History

The source of virtually all roadway air pollution emissions is the exhaust

The logical building block for this theory was the use of the Gaussian air pollutant dispersion equation for point sources. One of the early point source air pollutant plume dispersion equations was derived by Bosanquet and Pearson in 1936. Their equation did not include the effect of ground reflection of the pollutant plume. Sir Graham Sutton derived a point source air pollutant plume dispersion equation in 1947 which included the assumption of Gaussian distribution for the vertical and crosswind dispersion of the plume and also addressed the effect of ground reflection of the plume. Further advances were made by G. A. Briggs in model refinement and validation and by D.B. Turner for his user-friendly workbook that included screening calculations which do not require a computer.

In seeing the need to develop a line source model to approach the study of roadway air pollution, Michael Hogan and Richard Venti developed a closed form solution to integrating the point source equation in a series of publications.

While the ESL mathematical model was completed for a line source by 1970, model refinement resulted in a "strip source", emulating the horizontal extent of the roadway surface. This theory would be the precursor of area source dispersion models. But their focus was roadway simulation, so they proceeded with the development of a computer model by adding to the team Leda Patmore, a computer programmer in the field of atmospheric physics and satellite trajectory calculations. A working computer model was produced by late 1970; then the model was calibrated with carbon monoxide field measurements targeting from traffic on U.S. Route 101 in Sunnyvale, California.

The ESL model received endorsement from the U.S. Environmental Protection Agency (EPA) in the form of a major grant to validate the model using actual roadway tests of tracer gas sulfur hexafluoride dispersion. That gas was chosen since it does not occur naturally or in vehicular emissions and provides a unique tracer for such dispersion studies. Part of the Environmental Protection Agency's motives may have been to bring the model into public domain. After a successful validation through the EPA research, the model was soon put to use in a variety of settings to forecast air pollution levels in the vicinity of roadways. The ESL group applied the model to the U.S. Route 101 bypass project in Cloverdale, California, the extension of Interstate 66 through Arlington, Virginia, the widening of the New Jersey Turnpike through Raritan and East Brunswick, New Jersey, and several transportation projects in Boston for the Boston Transportation Planning Review.

By the early 1970s at least two other research groups were known to be actively developing some type to roadway air dispersion model: the Environmental Research and Technology group of Lexington, Massachusetts and Caltrans headquarters in Sacramento, California. The Caline model of Caltrans borrowed some of the technology from the ESL Inc. group, since Caltrans funded some of the early model application work in Cloverdale and other locations and was given rights to use parts of their model.

The theory

The resulting solution for an infinite line source is:

$$\chi = \int_0^\infty \frac{q}{\pi \left(ucdx^2 \right) \left(cos\alpha \right)} \left(\exp \frac{y^2}{2c^2x^2} \right) dx$$

where:

x is the distance from the observer to the roadway

y is the height of the observer

u is the mean wind speed

a is the angle of tilt of the line source relative to the reference frame

c and d are the standard deviation of horizontal and vertical wind directions (measured in radians) respectively.

This equation was integrated into a closed form solution using the error function (erf), and variations in geometry can be performed to include the full infinite line, line segment, elevated line, or arc made from segments. In any case one can calculate three-dimensional contours of resulting air pollutant concentrations and use the mathematical model to study alternative roadway designs, various assumptions of worst case meteorology or varying traffic conditions (for example, variations in truck mix, fleet emission controls, or vehicle speed).

The ESL research group also extended their model by introducing the area source concept of a vertical strip to simulate the mixing zone on the highway produced by vehicle turbulence. This model too was validated in 1971 and showed good correlation with field test data.

Example Applications of the Model

Roadway air dispersion modeling is also done for curved
roadways-North-South Express Highway, Malaysia

There were several early applications of the model in somewhat dramatic cases. In 1971 the Arlington Coalition on Transportation (ACT) was the plaintiff in an action against the Virginia Highway Commission over the extension of Interstate 66 through Arlington, Virginia, having filed a suit in the federal district court. The ESL model was used to produce calculations of air quality in the vicinity of the proposed highway. ACT won this case after a decision by the U.S. Fourth Circuit Court of Appeals. The court paid special attention to the plaintiff's expert calculations and testimony projecting that air quality levels would violate Federal ambient air quality standards as set forth in the Clean Air Act.

A second contentious case took place in East Brunswick, New Jersey where the New Jersey Turnpike Authority planned a major widening of the Turnpike. Again the roadway air dispersion model was employed to predict levels of air pollution for residences, schools and parks near the Turnpike. After an initial hearing in Superior Court where the ESL model results were set forth, the judge ordered the Turnpike Authority to negotiate with the plaintiff, Concerned Citizens of East Brunswick and develop air quality mitigation for the adverse effects. The Turnpike Authority hired ERT as its expert, and the two research teams negotiated a settlement to this case using the newly created roadway air dispersion models.

More Recent Model Refinements

The CALINE3 model is a steady-state Gaussian dispersion model designed to determine air pollution concentrations at receptor locations downwind of highways located in relatively uncomplicated terrain. CALINE3 is incorporated into the more elaborate CAL3QHC and CAL3QHCR models. CALINE3 is in widespread use due to its user friendly nature and promotion in governmental circles, but it falls short of analyzing the complexity of cases addressed by the original Hogan-Venti model. CAL3QHC and CAL3QHCR models are available in the Fortran programming language. They have options to model either particulate matter or carbon monoxide, and include algorithms to simulate queued traffic at signalized intersections .

In addition, several more recent models have been developed that employ non-steady state Lagrangian puff algorithms. The HYROAD dispersion model has been developed through the National Cooperative Highway Research Program's Project 25-06, incorporating ROADWAY-2 model puff and steady-state plume algorithms (Rao et al., 2002).

The TRAQSIM model, developed as part of a Ph.D dissertation with support by the U.S. Department of Transportation's Volpe National Transportation Systems Center's Air Quality Facility is currently under the care of Wyle. The model incorporates dynamic vehicle behavior with a non-steady state Gaussian puff algorithm. Unlike HYROAD, TRAQSIM combines traffic simulation, second-by-second modal emissions, and Gaussian puff dispersion into a fully integrated system (a true simulation) that models individual vehicles as discrete moving sources. TRAQSIM was developed as a next generation model to be the successor to the current CALINE3 and CAL3QHC regulatory models. The next step in the development of TRAQSIM is to incorporate methods to model the dispersion of particulate matter (PM) and hazardous air pollutants (HAPs).

Several models have been developed that handle complex urban meteorology resulting from urban canyons and highway configurations. The earliest such model development (1968-1970) was by the Air Pollution Control Office of the U.S. EPA in conjunction with New York City. The model was successfully applied to the Spadina Expressway in Toronto by Jack Fensterstock of the New York City Department of Air Resources,. Other examples include the Turner-Fairbank Highway Research Center's Canyon Plume Box model, now in version 3 (CPB-3), the National Environmental Research Institute of Denmark's Operational Street Pollution Model (OSPM), and the MICRO-CALGRID model, which includes photochemistry, allowing for both primary and secondary species to be modeled. Cornell University's CTAG model, which resolves vehicle-induced turbulence (VIT), road induced turbulence (RIT), chemical transformation and aerosol dynamics of air pollutants using turbulence reacting flow models. The CTAG model has also been applied to characterize highway-building environments and study effects of vegetation barriers on near-road air pollution.

Recent Applications in Legal Cases

Recent health literature indicating that residents near major roads face elevated rates of several adverse health outcomes has prompted legal dispute over the responsibility of transportation agencies to use roadway air dispersion models to characterize the impacts of new and expanded roadways, bus terminals, truck stops, and other sources.

Recently, the Sierra Club of Nevada sued the Nevada Department of Transportation and the Federal Highway Administration over its failure to assess the impact of the expansion of U.S. Route 95 in Las Vegas on neighborhood air quality. The Sierra Club asserted that a supplemental Environmental Impact Statement should be issued to address emissions of hazardous air pollutants and particulate matter from new motor vehicle traffic. The plaintiffs asserted that modeling tools were available, including the Environmental Protection Agency's MOBILE6.2 model, the CALINE3 dispersion model, and other relevant models. The defendants won in the U.S. District Court under Judge Philip Pro, who ruled that the transportation agencies had acted in a manner that was not "arbitrary and capricious," despite the agencies' technical arguments regarding the lack of available modeling tools being contradicted by a number of peer-reviewed studies published in scientific journals (e.g. Korenstein and Piazza, Journal of Environmental Health, 2002). On appeal to the U.S. Ninth Circuit, the Appeals Court stayed new construction on the highway pending the court's final decision. The Sierra Club and the defendants settled out of court, setting up a research program on the air quality impacts of U.S. Route 95 on nearby schools.

A number of other high-profile cases have prompted environmental groups to call for dispersion modeling to be used to assess the air quality impacts of new transportation projects on nearby communities, but to date state transportation agencies and the Federal Highways Administration has claimed that no tools are available, despite models and guidance available through EPA's Support Center for Regulatory Air Models (SCRAM).

Among the more contentious of cases the Detroit Intermodal Freight Terminal and Detroit River International Crossing (Michigan, USA), and the expansion of Interstate 70 East in Denver (Colorado, USA).

In all of these cases, community-based organizations have asserted that modeling tools are available, but transportation planning agencies have asserted that too much uncertainty exists in all of the steps. A major concern for community-based organizations has been transportation agencies' unwillingness to define the level of uncertainty that they are willing to tolerate in air quality analyses, how that compares to the Environmental Protection Agency's guideline on air quality models, which addresses uncertainty and accuracy in model use.

ADMS 3

The ADMS 3 (Atmospheric Dispersion Modelling System) is an advanced atmospheric pollution dispersion model for calculating concentrations of atmospheric pollutants emitted both continuously from point, line, volume and area sources, or intermittently from point sources. It was developed by Cambridge Environmental Research Consultants (CERC) of the UK in collaboration with the UK Meteorological Office, National Power plc (now INNOGY Holdings plc) and the University of Surrey. The first version of ADMS was released in 1993. The version of the ADMS model discussed on this page is version 3 and was released in February 1999. It runs on Microsoft Windows. The current release, ADMS 5 Service Pack 1, was released in April 2013 with a number of additional features.

Features and Capabilities of the ADMS 3

The model includes algorithms which take into account: downwash effects of nearby buildings within the path of the dispersing pollution plume; effects of complex terrain; effects of coastline locations; wet deposition, gravitational settling and dry deposition; short term fluctuations in pollutant concentration; chemical reactions; radioactive decay and gamma-dose; pollution plume rise as a function of distance; jets and directional releases; averaging time ranging from very short to annual; and condensed plume visibility. The system also includes a meteorological data input preprocessor.

The model is capable of simulating passive or buoyant continuous plumes as well as short duration puff releases. It characterizes the atmospheric turbulence by two parameters, the boundary layer depth and the Monin-Obukhov length, rather the single parameter Pasquill class.

ADMS 3 can simultaneously model up to 100 emission sources, of which:

- up to 100 may be point or jet sources

- up to 6 may be line, area or volume sources

- 1 may be a line source

The latest version (ADMS 5) allows up to 300 sources. Within that limit, up to 300 point sources, 30 line sources, 30 area sources and 30 volume sources may be modelled.

The performance of the model has been evaluated against various measured dispersion data sets.

Users of the ADMS 3

The users of ADMS 3 include:

- Governmental regulatory authorities including the UK Health and Safety Executive (HSE)

- Environment Agency of England and Wales

- Over 130 individual company licence holders in the UK

- Scottish Environment Protection Agency (SEPA) in Scotland

- Northern Ireland Environment Agency

- Governmental organisations including the Food Standards Agency (UK)

- Users in other European countries, Asia, Australia and the Middle East

- Accepted by the US Environmental Protection Agency as an "Alternative" model

AERMOD

The AERMOD atmospheric dispersion modeling system is an integrated system that includes three modules:

- A steady-state dispersion model designed for short-range (up to 50 kilometers) dispersion of air pollutant emissions from stationary industrial sources.

- A meteorological data preprocessor (AERMET) that accepts surface meteorological data, upper air soundings, and optionally, data from on-site instrument towers. It then calculates atmospheric parameters needed by the dispersion model, such as atmospheric turbulence characteristics, mixing heights, friction velocity, Monin-Obukov length and surface heat flux.

- A terrain preprocessor (AERMAP) whose main purpose is to provide a physical relationship between terrain features and the behavior of air pollution plumes. It generates location and height data for each receptor location. It also provides information that allows the dispersion model to simulate the effects of air flowing over hills or splitting to flow around hills.

AERMOD also includes PRIME (Plume Rise Model Enhancements) which is an algorithm for modeling the effects of downwash created by the pollution plume flowing over nearby buildings.

History of the Development of AERMOD

AERMOD was developed by the AERMIC (American Meteorological Society (AMS)/ United States Environmental Protection Agency (EPA) Regulatory Model Improvement Committee), a collaborative working group of scientists from the AMS and the EPA. The AERMIC was initially formed in 1991.

The AERMIC developed AERMOD in seven steps:

- Initial model formulation
- Developmental evaluation
- Internal peer review and beta testing
- Revised model formulation
- Performance evaluation and sensitivity testing
- External peer review
- Submission to the EPA for consideration as a regulatory model.

On April 21 of 2000, the EPA proposed that AERMOD be adopted as the EPA's preferred regulatory model for both simple and complex terrain. On November 9 of 2005, AERMOD was adopted by the EPA and promulgated as their preferred regulatory model, effective as of December 9 of 2005. The entire developmental and adoption process took 14 years (from 1991 to 2005).

Features and Capabilities of AERMOD

Some of the primary features and capabilities of AERMOD are:

- Source types: Multiple point, area and volume sources
- Source releases: Surface, near surface and elevated sources

- Source locations: Urban or rural locations. Urban effects are scaled by population.

- Plume types: Continuous, buoyant plumes

- Plume deposition: Dry or wet deposition of particulates and/or gases

- Plume dispersion treatment: Gaussian model treatment in horizontal and in vertical for stable atmospheres. Non-Gaussian treatment in vertical for unstable atmospheres

- Terrain types: Simple or complex terrain

- Building effects: Handled by PRIME downwash algorithms

- Meteorology data height levels: Accepts meteorology data from multiple heights

- Meteorological data profiles: Vertical profiles of wind, turbulence and temperature are created

CALPUFF

CALPUFF is an advanced, integrated Lagrangian puff modeling system for the simulation of atmospheric pollution dispersion distributed by the Atmospheric Studies Group at TRC Solutions.

It is maintained by the model developers and distributed by TRC. The model has been adopted by the United States Environmental Protection Agency (EPA) in its *Guideline on Air Quality Models* as a preferred model for assessing long range transport of pollutants and their impacts on Federal Class I areas and on a case-by-case basis for certain near-field applications involving complex meteorological conditions.

The integrated modeling system consists of three main components and a set of pre-processing and postprocessing programs. The main components of the modeling system are CALMET (a diagnostic 3-dimensional meteorological model), CALPUFF (an air quality dispersion model), and CALPOST (a postprocessing package). Each of these programs has a graphical user interface (GUI). In addition to these components, there are numerous other processors that may be used to prepare geophysical (land use and terrain) data in many standard formats, meteorological data (surface, upper air, precipitation, and buoy data), and interfaces to other models such as the Penn State/NCAR Mesoscale Model (MM5), the National Centers for Environmental Prediction (NCEP) Eta model and the RAMS meteorological model.

The CALPUFF model is designed to simulate the dispersion of buoyant, puff or continuous point and area pollution sources as well as the dispersion of buoyant, continuous line sources. The model also includes algorithms for handling the effect of downwash by nearby buildings in the path of the pollution plumes.

History

The CALPUFF model was originally developed by the Sigma Research Corporation (SRC) in the late 1980s under contract with the California Air Resources Board (CARB) and it was first issued in about 1990.

The Sigma Research Corporation subsequently became part of Earth Tech, Inc. After the US EPA designated CALPUFF as a preferred model in their *Guideline on Air Quality Models*, Earth Tech served as the designated distributor of the model.

In April 2006, ownership of the model switched from Earth Tech to the TRC Environmental Corporation. More recently ownership transferred to E^{x}ponent, who are currently (December 2015) responsible for maintaining and distributing the model.

FLACS

FLACS (FLame ACceleration Simulator) is a commercial Computational Fluid Dynamics (CFD) software used extensively for explosion modeling and atmospheric dispersion modeling within the field of industrial safety and risk assessment. Main application areas of FLACS are in petrochemical, process manufacturing, food processing, wood processing, metallurgical, and nuclear safety industries.

FLACS has dedicated modules to simulate gas explosion, dust explosion and explosions involving chemical explosives like TNT. FLACS is also extensively used to simulate flammable and toxic gas dispersion. It was applied in the investigation of many high profile accidents such as Buncefield fire, Piper Alpha, TWA Flight 800, and the Petrobras 36 platform.

History

FLACS software development started in-house in the early 1980s under the sponsorship program, Gas Explosion Safety (GSP), funded by the oil companies BP, Elf Aquitaine, Esso, Mobil, Norsk Hydro and Statoil. FLACS-86 was released to GSP sponsors in 1986. Continuous research and development from then onwards resulted in many commercial releases. In 2006, FLACS v8.1 was released to customers. Till then FLACS was developed for Unix and Linux platforms. In 2008, however, FLACS v9.0 was released for Microsoft Windows platform. FLACS v9.1 and FLACS-Wind was developed in 2010. A fully parallelized FLACSv10.0 (using OpenMP) with a new solver for incompressible flows was released in 2012. FLACSv10.0 also constitutes a Homogeneous Equilibrium Model (HEM) for two-phase flow calculations.

NAME (Dispersion Model)

The NAME atmospheric pollution dispersion model was first developed by the UK's Met Office in 1986 after the nuclear accident at Chernobyl, which demonstrated the need for a method that could predict the spread and deposition of radioactive gases or material released into the atmosphere.

The acronym, NAME, originally stood for the *Nuclear Accident ModEl*. The Met Office has revised and upgraded the model over the years and it is now used as a general purpose dispersion model. The current version is known as the NAME III (*Numerical Atmospheric dispersion Modeling Environment*) model. NAME III is currently operational and it will probably completely replace the original NAME model sometimes in 2006.

Features and Capabilities of NAME

NAME (in its current NAME III version) is a Lagrangian air pollution dispersion model for short range to global range scales. It employs 3-dimensional meteorological data provided by the Met Office's *Unified National Weather Prediction Model*. Random walk techniques using empirical turbulence profiles are utilized to represent turbulent mixing. In essence, NAME follows the 3-dimensional trajectories of parcels of the pollution plume and computes pollutant concentrations by Monte Carlo methods — that is, by direct simulation rather than solving equations.

NAME uses a puff technique when modelling dispersion over a short range which shortens the time needed to compute the pollutant concentrations at the receptors.

The model has the capability to calculate: the rise of buoyant plumes; deposition of pollution plume components due to rainfall (i.e., wet deposition); dry deposition; plume chemistry focusing on sulphate and nitrate chemistry; plume depletion via the decay of radioactive materials; the downwash effects of buildings.

The model can also be run 'backwards' to generate maps that locate possible plume originating sources.

The Met Office's Commitments to Emergency Response Service

The Met Office has international commitments to provide emergency response dispersion modelling services for releases of hazardous gases and materials into the atmosphere. Such events include the release of radioactive materials and emissions from erupting volcanoes. Those commitments are met by an operational group known as EMARC who are supported by a Met Office team of dispersion modelling staff. That team is also responsible for the scientific development of NAME III which, combined with the Met Office numerical weather prediction model, is used to provide the dispersion modelling services needed to implement the listed commitments:

- The WMO (*World Meteorological Office*) has designated the Met Office to operate one of the worldwide RSMCs (*Regional Specialist Meteorological Centre*),
which the Met Office located at Bracknell.

- The Met Office has also been designated a VAAC (*Volcanic Ash Advisory Centre*) which is part of the IAVW (*International Airways Volcano Watch*) set up
by the ICAO (*International Civil Aviation Organization*).

Over the years, NAME has been applied to radioactive releases, the Kuwaiti oil fires,
major industrial fires and chemical spills, and two volcanic eruptions in Iceland.

References

- Beychok, Milton R. (2005). Fundamentals Of Stack Gas Dispersion (4th ed.). author-published.
ISBN 0-9644588-0-2.

- Turner, D.B. (1994). Workbook of atmospheric dispersion estimates: an introduction to dispersion
modeling (2nd ed.). CRC Press. ISBN 1-56670-023-X.

- Beychok, M.R. (2005). Fundamentals Of Stack Gas Dispersion (4th ed.). author-published.
ISBN 0-9644588-0-2.

Indoor Air Quality: Assessment and Control

Indoor air quality or IAQ refers to the quality of the air within buildings. The pollution that is caused indoors is a major health hazard in a number of countries; one of the major causes of air pollution is the burning of biomass. Passive smoking, HVAC, air conditioning and air handle are some of the reasons for varied indoor air quality. The major categories of indoor air quality are dealt with great details in the chapter.

Indoor Air Quality

Indoor air quality (IAQ) is a term which refers to the air quality within and around buildings and structures, especially as it relates to the health and comfort of building occupants. IAQ can be affected by gases (including carbon monoxide, radon, volatile organic compounds), particulates, microbial contaminants (mold, bacteria), or any mass or energy stressor that can induce adverse health conditions. Source control, filtration and the use of ventilation to dilute contaminants are the primary methods for improving indoor air quality in most buildings. Residential units can further improve indoor air quality by routine cleaning of carpets and area rugs.

A common air filter, being cleaned with a vacuum cleaner

Determination of IAQ involves the collection of air samples, monitoring human exposure to pollutants, collection of samples on building surfaces, and computer modelling of air flow inside buildings.

IAQ is part of indoor environmental quality (IEQ), which includes IAQ as well as other physical and psychological aspects of life indoors (e.g., lighting, visual quality, acoustics, and thermal comfort).

Indoor air pollution in developing nations is a major health hazard. A major source of indoor air pollution in developing countries is the burning of biomass (e.g. wood, charcoal, dung, or crop residue) for heating and cooking. The resulting exposure to high levels of particulate matter resulted in between 1.5 million and 2 million deaths in 2000.

Common Pollutants

Second-hand Smoke

Second-hand smoke is tobacco smoke which affects other people other than the 'active' smoker. Second-hand tobacco smoke includes both a gaseous and a particulate phase, with particular hazards arising from levels of carbon monoxide (as indicated below) and very small particulates (at PM2.5 size) which get past the lung's natural defenses. The only certain method to improve indoor air quality as regards second-hand smoke is the implementation of comprehensive smoke-free laws.

Radon

Radon is an invisible, radioactive atomic gas that results from the radioactive decay of radium, which may be found in rock formations beneath buildings or in certain building materials themselves. Radon is probably the most pervasive serious hazard for indoor air in the United States and Europe, probably responsible for tens of thousands of deaths from lung cancer each year. There are relatively simple test kits for do-it-yourself radon gas testing, but if a home is for sale the testing must be done by licensed person in some U.S. states. Radon gas enters buildings as a soil gas and is a heavy gas and thus will tend to accumulate at the lowest level. Radon may also be introduced into a building through drinking water particularly from bathroom showers. Building materials can be a rare source of radon, but little testing is carried out for stone, rock or tile products brought into building sites; radon accumulation is greatest for well insulated homes. The half life for radon is 3.8 days, indicating that once the source is removed, the hazard will be greatly reduced within a few weeks. Radon mitigation methods include sealing concrete slab floors, basement foundations, water drainage systems, or by increasing ventilation. They are usually cost effective and can greatly reduce or even eliminate the contamination and the associated health risks.

Molds and other Allergens

These biological chemicals can arise from a host of means, but there are two common classes: (a) moisture induced growth of mold colonies and (b) natural substances released into the air such as animal dander and plant pollen. Mold is always associated with moisture, and its growth can be inhibited by keeping humidity levels below 50%. Moisture buildup inside buildings may arise from water penetrating compromised areas of the building envelope or skin, from plumbing leaks, from condensation due to improper ventilation, or from ground moisture penetrating a building part. In areas where cellulosic materials (paper and wood, including drywall) become moist and fail to dry within 48 hours, mold mildew can propagate and release allergenic spores into the air.

In many cases, if materials have failed to dry out several days after the suspected water event, mold growth is suspected within wall cavities even if it is not immediately visible. Through a mold investigation, which may include destructive inspection, one should be able to determine the presence or absence of mold. In a situation where there is visible mold and the indoor air quality may have been compromised, mold remediation may be needed. Mold testing and inspections should be carried out by an independent investigator to avoid any conflict of interest and to insure accurate results; free mold testing offered by remediation companies is not recommended.

There are some varieties of mold that contain toxic compounds (mycotoxins). However, exposure to hazardous levels of mycotoxin via inhalation is not possible in most cases, as toxins are produced by the fungal body and are not at significant levels in the released spores. The primary hazard of mold growth, as it relates to indoor air quality, comes from the allergenic properties of the spore cell wall. More serious than most allergenic properties is the ability of mold to trigger episodes in persons that already have asthma, a serious respiratory disease.

Carbon Monoxide

One of the most acutely toxic indoor air contaminants is carbon monoxide (CO), a colourless, odourless gas that is a byproduct of incomplete combustion of fossil fuels. Common sources of carbon monoxide are tobacco smoke, space heaters using fossil fuels, defective central heating furnaces and automobile exhaust. By depriving the brain of oxygen, high levels of carbon monoxide can lead to nausea, unconsciousness and death. According to the American Conference of Governmental Industrial Hygienists (ACGIH), the time-weighted average (TWA) limit for carbon monoxide (630-08-0) is 25 ppm.

Indoor levels of CO are systematically improving due to increasing implementation of smoke-free laws.

Volatile Organic Compounds

Volatile organic compounds (VOCs) are emitted as gases from certain solids or liquids. VOCs include a variety of chemicals, some of which may have short- and long-term adverse health effects. Concentrations of many VOCs are consistently higher indoors (up to ten times higher) than outdoors. VOCs are emitted by a wide array of products numbering in the thousands. Examples include: paints and lacquers, paint strippers, cleaning supplies, pesticides, building materials and furnishings, office equipment such as copiers and printers, correction fluids and carbonless copy paper, graphics and craft materials including glues and adhesives, permanent markers, and photographic solutions.

Chlorinated drinking water releases chloroform when hot water is used in the home. Benzene is emitted from fuel stored in attached garages. Overheated cooking oils emit acrolein and formaldehyde. A meta-analysis of 77 surveys of VOCs in homes in the US found the top ten riskiest indoor air VOCs were acrolein, formaldehyde, benzene, hexachlorobutadiene, acetaldehyde, 1,3-butadiene, benzyl chloride, 1,4-dichloroben-zene, carbon tetrachloride, acrylonitrile, and vinyl chloride. These compounds exceed-ed health standards in most homes.

Organic chemicals are widely used as ihgredients in household products. Paints, var-nishes, and wax all contain organic solvents, as do many cleaning, disinfecting, cos-metic, degreasing, and hobby products. Fuels are made up of organic chemicals. All of these products can release organic compounds during usage, and, to some degree, when they are stored. Testing emissions from building materials used indoors has be-come increasingly common for floor coverings, paints, and many other important in-door building materials and finishes.

Several initiatives envisage to reduce indoor air contamination by limiting VOC emis-sions from products. There are regulations in France and in Germany, and numerous voluntary ecolabels and rating systems containing low VOC emissions criteria such as EMICODE, M1, Blue Angel and Indoor Air Comfort in Europe, as well as Califor-nia Standard CDPH Section 01350 and several others in the USA. These initiatives changed the marketplace where an increasing number of low-emitting products has become available during the last decades.

At least 18 Microbial VOCs (MVOCs) have been characterised including 1-octen-3-ol, 3-methylfuran, 2-pentanol, 2-hexanone, 2-heptanone, 3-octanone, 3-octanol, 2-octen-1-ol, 1-octene, 2-pentanone, 2-nonanone, borneol, geosmin, 1-butanol, 3-methyl-1-bu-tanol, 3-methyl-2-butanol, and thujopsene. The first of these compounds is called mushroom alcohol. The last four are products of *Stachybotrys chartarum*, which has been linked with sick building syndrome.

Legionella

Legionellosis or Legionnaire's Disease is caused by a waterborne bacterium *Legionella* that grows best in slow-moving or still, warm water. The primary route of exposure is through the creation of an aerosol effect, most commonly from evaporative cooling towers or show-erheads. A common source of Legionella in commercial buildings is from poorly placed or maintained evaporative cooling towers, which often release water in an aerosol which may enter nearby ventilation intakes. Outbreaks in medical facilities and nursing homes, where patients are immuno-suppressed and immuno-weak, are the most commonly re-ported cases of Legionellosis. More than one case has involved outdoor fountains in public attractions. The presence of Legionella in commercial building water supplies is highly un-der-reported, as healthy people require heavy exposure to acquire infection.

Legionella testing typically involves collecting water samples and surface swabs from evaporative cooling basins, shower heads, faucets/taps, and other locations where warm water collects. The samples are then cultured and colony forming units (cfu) of Legionella are quantified as cfu/Liter.

Legionella is a parasite of protozoans such as amoeba, and thus requires conditions suitable for both organisms. The bacterium forms a biofilm which is resistant to chem-ical and antimicrobial treatments, including chlorine. Remediation for Legionella out-breaks in commercial buildings vary, but often include very hot water flushes (160 °F; 70 °C), sterilisation of standing water in evaporative cooling basins, replacement of shower heads, and in some cases flushes of heavy metal salts. Preventative measures include adjusting normal hot water levels to allow for 120 °F at the tap, evaluating facility design layout, removing faucet aerators, and periodic testing in suspect areas.

Other Bacteria

There are many bacteria of health significance found in indoor air and on indoor surfac-es. The role of microbes in the indoor environment is increasingly studied using mod-ern gene-based analysis of environmental samples. Currently efforts are under way to link microbial ecologists and indoor air scientists to forge new methods for analysis and to better interpret the results.

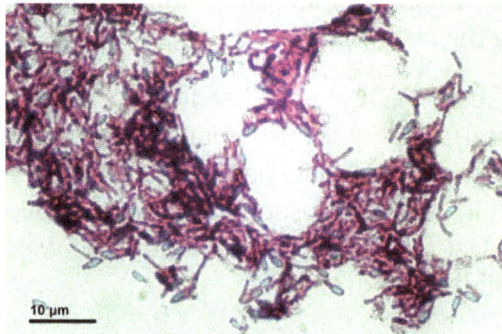

Bacteria (26 2 27) Airborne microbes

"There are approximately ten times as many bacterial cells in the human flora as there are human cells in the body, with large numbers of bacteria on the skin and as gut flo-ra." A large fraction of the bacteria found in indoor air and dust are shed from humans. Among the most important bacteria known to occur in indoor air are Mycobacterium tuberculosis, Staphylococcus aureus, Streptococcus pneumoniae.

Asbestos Fibers

Many common building materials used before 1975 contain asbestos, such as some floor tiles, ceiling tiles, shingles, fireproofing, heating systems, pipe wrap, taping muds, mastics, and other insulation materials. Normally, significant releases of asbestos fiber do not occur unless the building materials are disturbed, such as by cutting, sanding, drilling, or building remodelling. Removal of asbestos-containing materials is not always optimal because the fibers can be spread into the air during the removal process. A management program for intact asbestos-containing materials is often recommended instead.

When asbestos-containing material is damaged or disintegrates, microscopic fibers are dispersed into the air. Inhalation of asbestos fibers over long exposure times is associated with increased incidence of lung cancer, in particular the specific form mesothelioma. The risk of lung cancer from inhaling asbestos fibers is also greater to smokers. The symptoms of the disease do not usually appear until about 20 to 30 years after the first exposure to asbestos.

Asbestos is found in older homes and buildings, but occurs most commonly in schools and industrial settings. The US Federal Government (www.osha.gov) and some states have set standards for acceptable levels of asbestos fibers in indoor air. There are particularly stringent regulations applicable to schools.

Carbon Dioxide

Carbon dioxide (CO_2) is a relatively easy to measure surrogate for indoor pollutants emitted by humans, and correlates with human metabolic activity. Carbon dioxide at levels that are unusually high indoors may cause occupants to grow drowsy, to get headaches, or to function at lower activity levels. Humans are the main indoor source of carbon dioxide in most buildings. Indoor CO_2 levels are an indicator of the adequacy of outdoor air ventilation relative to indoor occupant density and metabolic activity.

To eliminate most complaints, the total indoor CO_2 level should be reduced to a difference of less than 600 ppm above outdoor levels. The National Institute for Occupational Safety and Health (NIOSH) considers that indoor air concentrations of carbon dioxide that exceed 1,000 ppm are a marker suggesting inadequate ventilation. The UK standards for schools say that carbon dioxide in all teaching and learning spaces, when measured at seated head height and averaged over the whole day should not exceed 1,500 ppm. The whole day refers to normal school hours (i.e. 9:00am to 3:30pm) and includes unoccupied periods such as lunch breaks. In Hong Kong, the EPD established indoor air quality objectives for office buildings and public places in which a carbon dioxide level below 1,000 ppm is considered to be good. European standards limit carbon dioxide to 3,500 ppm. OSHA limits carbon dioxide concentration in the workplace to 5,000 ppm for prolonged periods, and 35,000 ppm for 15 minutes. These higher limits are concerned with avoiding loss of consciousness (fainting), and do not address impaired cognitive performance and energy, which begin to occur at lower concentrations of carbon dioxide.

Carbon dioxide concentrations increase as a result of human occupancy, but lag in time behind cumulative occupancy and intake of fresh air. The lower the air exchange rate, the slower the buildup of carbon dioxide to quasi "steady state" concentrations on which the NIOSH and UK guidance are based. Therefore, measurements of carbon dioxide for purposes of assessing the adequacy of ventilation need to be made after an extended period of steady occupancy and ventilation - in schools at least 2 hours, and in offices at least 3 hours - for concentrations to be a reasonable indicator of ventilation adequacy. Portable instruments used to measure carbon dioxide should be calibrated frequently, and outdoor measurements used for calculations should be made close in time to indoor measurements. Corrections for temperature effects on measurements made outdoors may also be necessary.

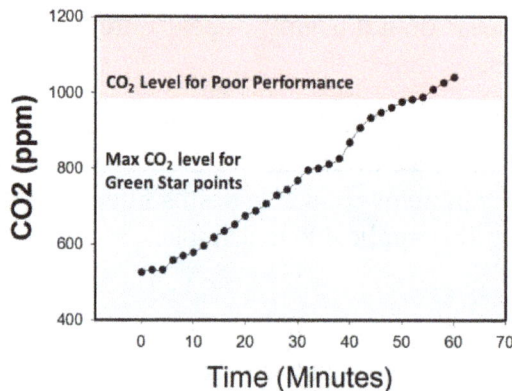

CO2 levels in an enclosed office room can increase to over 1,000 ppm within 45 minutes.

Carbon dioxide concentrations in closed or confined rooms can increase to 1,000 ppm within 45 minutes of enclosure. For example, in a 3.5-by-4-metre (11 ft × 13 ft) sized office, atmospheric carbon dioxide increased from 500 ppm to over 1,000 ppm within 45 minutes of ventilation cessation and closure of windows and doors.

Ozone

Ozone is produced by ultraviolet light from the Sun hitting the Earth's atmosphere (especially in the ozone layer), lightning, certain high-voltage electric devices (such as air ionizers), and as a by-product of other types of pollution.

Ozone exists in greater concentrations at altitudes commonly flown by passenger jets. Reactions between ozone and onboard substances, including skin oils and cosmetics, can produce toxic chemicals as by-products. Ozone itself is also irritating to lung tissue and harmful to human health. Larger jets have ozone filters to reduce the cabin concentration to safer and more comfortable levels.

Outdoor air used for ventilation may have sufficient ozone to react with common indoor pollutants as well as skin oils and other common indoor air chemicals or surfaces. Particular concern is warranted when using "green" cleaning products based on citrus or terpene extracts, because these chemicals react very quickly with ozone to form toxic and irritating chemicals as well as fine and ultrafine particles. Ventilation with outdoor air containing elevated ozone concentrations may complicate remediation attempts.

Prompt Cognitive Deficits

In 2015, experimental studies reported the detection of significant episodic (situational) cognitive impairment from impurities in the air breathed by test subjects who were not informed about changes in the air quality. Researchers at the Harvard University and SUNY Upstate Medical University and Syracuse University measured the cognitive performance of 24 participants in three different controlled laboratory atmospheres that simulated those found in "conventional" and "green" buildings, as well as green buildings with enhanced ventilation. Performance was evaluated objectively using the widely used Strategic Management Simulation software simulation tool, which is a well-validated assessment test for executive decision-making in an unconstrained situation allowing initiative and improvisation. Significant deficits were observed in the performance scores achieved in increasing concentrations of either volatile organic compounds (VOCs) or carbon dioxide, while keeping other factors constant. The highest impurity levels reached are not uncommon in some classroom or office environments.

Effect of Indoor Plants

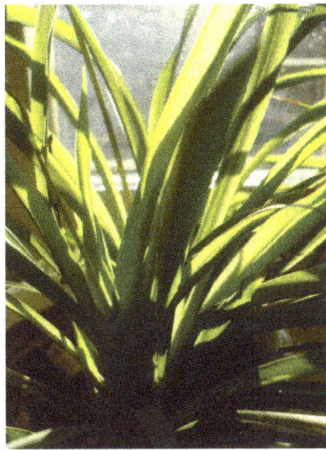

Spider plants *(Chlorophytum comosum)* absorb some airborne contaminants

Houseplants together with the medium in which they are grown can reduce components of indoor air pollution, particularly volatile organic compounds (VOC) such as benzene, toluene, and xylene. Plants remove CO_2 and release oxygen and water, although the quantitative impact for house plants is small. Most of the effect is attributed to the growing medium alone, but even this effect has finite limits associated with the type and quantity of medium and the flow of air through the medium. The effect of house plants on VOC concentrations was investigated in one study, done in a static

chamber, by NASA for possible use in space colonies. The results showed that the removal of the challenge chemicals was roughly equivalent to that provided by the ventilation that occurred in a very energy efficient dwelling with a very low ventilation rate,
an air exchange rate of about 1/10 per hour. Therefore, air leakage in most homes, and
in non-residential buildings too, will generally remove the chemicals faster than the researchers reported for the plants tested by NASA. The most effective household plants
reportedly included aloe vera, English ivy, and Boston fern for removing chemicals and
biological compounds.

Plants also appear to reduce airborne microbes, molds, and increase humidity. However, the increased humidity can itself lead to increased levels of mold and even
VOCs.

When CO_2 concentrations are elevated indoors relative to outdoor concentrations, it
is only an indicator that ventilation is inadequate to remove metabolic products associated with human occupancy. Plants require CO_2 to grow and release oxygen when
they consume CO_2. A study published in the journal *Environmental Science & Technology* considered uptake rates of ketones and aldehydes by the peace lily (Spathiphyl-
lum clevelandii) and golden pothos (Epipremnum aureum.) Akira Tani and C. Nicholas
Hewitt found "Longer-term fumigation results revealed that the total uptake amounts
were 30–100 times as much as the amounts dissolved in the leaf, suggesting that volatile organic carbons are metabolized in the leaf and/or translocated through the pet-
iole." It is worth noting the researchers sealed the plants in Teflon bags. "No VOC loss
was detected from the bag when the plants were absent. However, when the plants
were in the bag, the levels of aldehydes and ketones both decreased slowly but continuously, indicating removal by the plants". Studies done in sealed bags do not faithfully
reproduce the conditions in the indoor environments of interest. Dynamic conditions
with outdoor air ventilation and the processes related to the surfaces of the building
itself and its contents as well as the occupants need to be studied.

While results do indicate house plants may be effective at removing some VOCs from
air supplies, a review of studies between 1989 and 2006 on the performance of houseplants as air cleaners, presented at the Healthy Buildings 2009 conference in Syracuse,
NY, concluded "...indoor plants have little, if any, benefit for removing indoor air of
VOC in residential and commercial buildings."

Since high humidity is associated with increased mold growth, allergic responses, and
respiratory responses, the presence of additional moisture from houseplants may not
be desirable in all indoor settings.

HVAC Design

Environmentally sustainable design concepts also include aspects related to the commercial and residential heating, ventilation and air-conditioning (HVAC) industry.

Among several considerations, one of the topics attended to is the issue of indoor air quality throughout the design and construction stages of a building's life.

One technique to reduce energy consumption while maintaining adequate air quality, is demand controlled ventilation. Instead of setting throughput at a fixed air replacement rate, carbon dioxide sensors are used to control the rate dynamically, based on the emissions of actual building occupants.

For the past several years, there have been many debates among indoor air quality specialists about the proper definition of indoor air quality and specifically what constitutes "acceptable" indoor air quality.

One way of quantitatively ensuring the health of indoor air is by the frequency of effective turnover of interior air by replacement with outside air. In the UK, for example, classrooms are required to have 2.5 outdoor air changes per hour. In halls, gym, dining, and physiotherapy spaces, the ventilation should be sufficient to limit carbon dioxide to 1,500 ppm. In the USA, and according to ASHRAE Standards, ventilation in classrooms is based on the amount of outdoor air per occupant plus the amount of outdoor air per unit of floor area, not air changes per hour. Since carbon dioxide indoors comes from occupants and outdoor air, the adequacy of ventilation per occupant is indicated by the concentration indoors minus the concentration outdoors. The value of 615 ppm above the outdoor concentration indicates approximately 15 cubic feet per minute of outdoor air per adult occupant doing sedentary office work where outdoor air contains 385 ppm, the current global average atmospheric CO_2 concentration. In classrooms, the requirements in the ASHRAE standard 62.1, Ventilation for Acceptable Indoor Air Quality, would typically result in about 3 air changes per hour, depending on the occupant density. Of course the occupants aren't the only source of pollutants, so outdoor air ventilation may need to be higher when unusual or strong sources of pollution exist indoors. When outdoor air is polluted, then bringing in more outdoor air can actually worsen the overall quality of the indoor air and exacerbate some occupant symptoms related to outdoor air pollution. Generally, outdoor country air is better than indoor city air. Exhaust gas leakages can occur from furnace metal exhaust pipes that lead to the chimney when there are leaks in the pipe and the pipe gas flow area diameter has been reduced.

The use of air filters can trap some of the air pollutants. The Department of Energy's Energy Efficiency and Renewable Energy section wrote "[Air] Filtration should have a Minimum Efficiency Reporting Value (MERV) of 13 as determined by ASHRAE 52.2-1999." Air filters are used to reduce the amount of dust that reaches the wet coils. Dust can serve as food to grow molds on the wet coils and ducts and can reduce the efficiency of the coils.

Moisture management and humidity control requires operating HVAC systems as designed. Moisture management and humidity control may conflict with efforts to try to optimize the operation to conserve energy. For example, Moisture management and humidity control requires systems to be set to supply Make Up Air at lower temperatures (design levels), instead of the higher temperatures sometimes used to conserve energy in cooling-dominated climate conditions. However, for most of the US and many parts of Europe and Japan, during the majority of hours of the year, outdoor air temperatures are cool enough that the air does not need further cooling to provide thermal comfort indoors. However, high humidity outdoors creates the need for careful attention to humidity levels indoors. High humidities give rise to mold growth and moisture indoors is associated with a higher prevalence of occupant respiratory problems.

The "dew point temperature" is an absolute measure of the moisture in air. Some facilities are being designed with the design dew points in the lower 50s °F, and some in the upper and lower 40s °F. Some facilities are being designed using desiccant wheels with gas fired heater to dry out the wheel enough to get the required dew points. On those systems, after the moisture is removed from the make up air, a cooling coil is used to lower the temperature to the desired level.

Commercial buildings, and sometimes residential, are often kept under slightly positive air pressure relative to the outdoors to reduce infiltration. Limiting infiltration helps with moisture management and humidity control.

Dilution of indoor pollutants with outdoor air is effective to the extent that outdoor air is free of harmful pollutants. Ozone in outdoor air occurs indoors at reduced concentrations because ozone is highly reactive with many chemicals found indoors. The products of the reactions between ozone and many common indoor pollutants include organic compounds that may be more odorous, irritating, or toxic than those from which they are formed. These products of ozone chemistry include formaldehyde, higher molecular weight aldehydes, acidic aerosols, and fine and ultrafine particles, among others. The higher the outdoor ventilation rate, the higher the indoor ozone concentration and the more likely the reactions will occur, but even at low levels, the reactions will take place. This suggests that ozone should be removed from ventilation air, especially in areas where outdoor ozone levels are frequently high. Recent research has shown that mortality and morbidity increase in the general population during periods of higher outdoor ozone and that the threshold for this effect is around 20 parts per billion (ppb).

Building Ecology

It is common to assume that buildings are simply inanimate physical entities, relatively stable over time. This implies that there is little interaction between the triad of the building, what is in it (occupants and contents), and what is around it (the larger environment). We commonly see the overwhelming majority of the mass of material in a building as relatively unchanged physical material over time. In fact, the true nature of buildings can be viewed as the result of a complex set of dynamic interactions among their physical, chemical, and biological dimensions. Buildings can be described and understood as complex systems. Research applying the approaches ecologists use to the understanding of ecosystems can help increase our understanding. "Building ecology " is proposed here as the application of those approaches to the built environment considering the dynamic system of buildings, their occupants, and the larger environment.

Buildings constantly evolve as a result of the changes in the environment around them as well as the occupants, materials, and activities within them. The various surfaces and the air inside a building are constantly interacting, and this interaction results in changes in each. For example, we may see a window as changing slightly over time as it becomes dirty, then is cleaned, accumulates dirt again, is cleaned again, and so on through its life. In fact, the "dirt" we see may be evolving as a result of the interactions among the moisture, chemicals, and biological materials found there.

Buildings are designed or intended to respond actively to some of these changes in and around them with heating, cooling, ventilating, air cleaning or illuminating systems. We clean, sanitize, and maintain surfaces to enhance their appearance, performance, or longevity. In other cases, such changes subtly or even dramatically alter buildings in ways that may be important to their own integrity or their impact on building occupants through the evolution of the physical, chemical, and biological processes that define them at any time. We may find it useful to combine the tools of the physical sciences with those of the biological sciences and, especially, some of the approaches used by scientists studying ecosystems, in order to gain an enhanced understanding of the environments in which we spend the majority of our time, our buildings.

Building ecology was first described by Hal Levin in an article in the April 1981 issue of Progressive Architecture magazine. A longer discussion of Building ecology can be found at and extensive resources can be found on the Building Ecology web site Building ecology.com.

Institutional Programs

The topic of IAQ has become popular due to the greater awareness of health problems caused by mold and triggers to asthma and allergies. In the US, awareness has also been increased by the involvement of the United States Environmental Protection Agency, who have developed an "IAQ Tools for Schools" program to help improve the indoor environmental conditions in educational institutions. The National Institute for Occupational Safety and Health conducts Health Hazard Evaluations (HHEs) in workplaces at the request of employees, authorised representative of employees, or employers, to determine whether any substance normally found in the place of employment has potentially toxic effects, including indoor air quality.

A variety of scientists work in the field of indoor air quality including chemists, physicists, mechanical engineers, biologists, bacteriologists and computer scientists. Some of these professionals are certified by organisations such as the American Industrial

Hygiene Association, the American Indoor Air Quality Council and the Indoor Environmental Air Quality Council.

On the international level, the International Society of Indoor Air Quality and Climate (ISIAQ), formed in 1991, organises two major conferences, the Indoor Air and the Healthy Buildings series. ISIAQ's journal *Indoor Air* is published 6 times a year and contains peer-reviewed scientific papers with an emphasis on interdisciplinary studies including exposure measurements, modeling, and health outcomes.

Passive Smoking

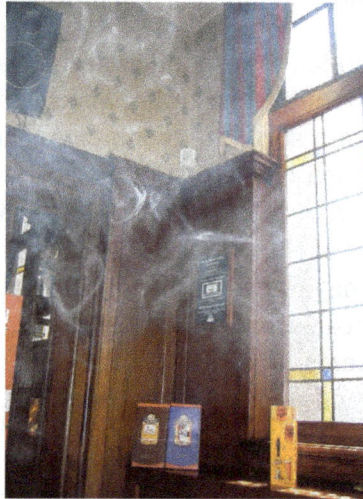

Tobacco smoke in an Irish pub before a smoking ban came into effect on March 29, 2004

Passive smoking is the inhalation of smoke, called second-hand smoke (SHS), or environmental tobacco smoke (ETS), by persons other than the intended "active" smoker. It occurs when tobacco smoke permeates any environment, causing its inhalation by people within that environment. Exposure to second-hand tobacco smoke causes disease, disability, and death. The health risks of second-hand smoke are a matter of scientific consensus. These risks have been a major motivation for smoke-free laws in workplaces and indoor public places, including restaurants, bars and night clubs, as well as some open public spaces.

Concerns around second-hand smoke have played a central role in the debate over the harms and regulation of tobacco products. Since the early 1970s, the tobacco industry has viewed public concern over second-hand smoke as a serious threat to its business interests. Harm to bystanders was perceived as a motivator for stricter regulation of tobacco products. Despite the industry's awareness of the harms of second-hand smoke as early as the 1980s, the tobacco industry coordinated a scientific controversy with the aim of forestalling regulation of their products.

Effects

Second-hand smoke causes many of the same diseases as direct smoking, including cardiovascular diseases, lung cancer, and respiratory diseases. These diseases include:

- Cancer:

    o General: overall increased risk; reviewing the evidence accumulated on a worldwide basis, the International Agency for Research on Cancer concluded in 2004 that "Involuntary smoking (exposure to secondhand or 'environmental' tobacco smoke) is carcinogenic to humans."

    o Lung cancer: passive smoking is a risk factor for lung cancer. In the United States passive smoke is estimated to cause more than 7,000 deaths from lung cancer a year among non-smokers.

    o Breast cancer: The California Environmental Protection Agency concluded in 2005 that passive smoking increases the risk of breast cancer in younger, primarily premenopausal women by 70% and the US Surgeon General has concluded that the evidence is "suggestive," but still insufficient to assert such a causal relationship. In contrast, the International Agency for Research on Cancer concluded in 2004 that there was "no support for a causal relation between involuntary exposure to tobacco smoke and breast cancer in never-smokers." A 2015 meta-analysis found that the evidence that passive smoking moderately increased the risk of breast cancer had become "more substantial than a few years ago."

    o Pancreatic cancer: A 2012 meta-analysis found no evidence that passive smoking was associated with an increased risk of pancreatic cancer.

    o Cervical cancer: A 2015 overview of systematic reviews found that exposure to second-hand smoke increased the risk of cervical cancer.

- Circulatory system: risk of heart disease, reduced heart rate variability.

    o Epidemiological studies have shown that both active and passive cigarette smoking increase the risk of atherosclerosis.

    o Passive smoking is strongly associated with an increased risk of stroke, and this increased risk is disproportionately high at low levels of exposure.

- Lung problems:

    o Risk of asthma.

    o Risk of chronic obstructive pulmonary disease (COPD)

  o According to a 2015 review, passive smoking may increase the risk of tuberculosis infection and accelerate the progression of the disease, but the evidence remains weak.

- Cognitive impairment and dementia: Exposure to secondhand smoke may increase the risk of cognitive impairment and dementia in adults 50 and over.

- Mental health: Exposure to secondhand smoke is associated with an increased risk of depressive symptoms.

- During pregnancy:

  o Low birth weight.

  o Premature birth (Note that evidence of the causal link is only described as "suggestive" by the US Surgeon General in his 2006 report.) Laws limiting smoking decrease premature births.

  o Stillbirth and congenital malformations in children

  o Recent studies comparing women exposed to Environmental Tobacco Smoke and non-exposed women, demonstrate that women exposed while pregnant have higher risks of delivering a child with congenital abnormalities, longer lengths, smaller head circumferences, and low birth weight.

- General:

  o Worsening of asthma, allergies, and other conditions.

  o Type 2 diabetes. It remains unclear whether the association between passive smoking and diabetes is causal.

- Risk of carrying Neisseria meningitidis or Streptococcus pneumoniae.

- Overall increased risk of death in both adults, where it is estimated to kill 53,000 nonsmokers per year, making it the 3rd leading cause of preventable death in the U.S, and in children. The World Health Organization states that passive smoking causes about 600,000 deaths a year, and about 1% of the global burden of disease.

Risk to Children

- Sudden infant death syndrome (SIDS). In his 2006 report, the US Surgeon General concludes: "The evidence is sufficient to infer a causal relationship between exposure to secondhand smoke and sudden infant death syndrome." Secondhand smoking has been estimated to be associated with 430 SIDS deaths in the United States annually.

- Asthma

- Lung infections, also including more severe illness with bronchiolitis and bronchitis, and worse outcome, as well as increased risk of developing tuberculosis if exposed to a carrier. In the United States, it is estimated that second-hand smoke has been associated with between 150,000 and 300,000 lower respiratory tract infections in infants and children under 18 months of age, resulting in between 7,500 and 15,000 hospitalizations each year.

- Impaired respiratory function and slowed lung growth

- Allergies

- Maternal passive smoking increases the risk of non-syndromic orofacial clefts by 50% among their children.

- Prenatal and childhood passive smoke exposure does not appear to increase the risk of inflammatory bowel disease.

- Learning difficulties, developmental delays, executive function problems, and neurobehavioral effects. Animal models suggest a role for nicotine and carbon monoxide in neurocognitive problems.

- An increase in tooth decay (as well as related salivary biomarkers) has been associated with passive smoking in children.

- Increased risk of middle ear infections.

- Invasive meningococcal disease.

- Maternal exposure to secondhand smoke exposure during pregnancy is associated with an increased risk of neural tube defects.

- Miscarriage: a 2014 meta-analysis found that maternal secondhand smoke exposure increased the risk of miscarriage by 11%.

Evidence

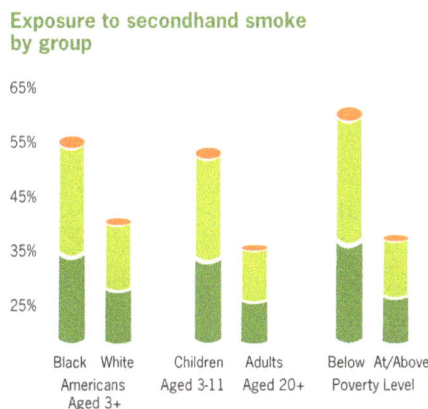

Exposure to secondhand smoke by age, race, and poverty level in the US.

Epidemiological studies show that non-smokers exposed to second-hand smoke are at risk for many of the health problems associated with direct smoking. Most of the research has come from studies of nonsmokers who are married to a smoker. Those conclusions are also backed up by further studies of workplace exposure to smoke.

In 1992, a review estimated that second-hand smoke exposure was responsible for 35,000 to 40,000 deaths per year in the United States in the early 1980s. The absolute risk increase of heart disease due to ETS was 2.2%, while the attributable risk percent was 23%. A 1997 meta-analysis found that second-hand smoke exposure increased the risk of heart disease by a quarter, and two 1999 meta-analyses reached similar conclusions.

Evidence shows that inhaled sidestream smoke, the main component of second-hand smoke, is about four times more toxic than mainstream smoke. This fact has been known to the tobacco industry since the 1980s, though it kept its findings secret. Some scientists believe that the risk of passive smoking, in particular the risk of developing coronary heart diseases, may have been substantially underestimated.

In 1997, a meta-analysis on the relationship between secondhand smoke exposure and lung cancer concluded that such exposure caused lung cancer. The increase in risk was estimated to be 24 percent among non-smokers who lived with a smoker. In 2000, Copas and Shi reported that there was clear evidence of publication bias in the studies included in this meta-analysis. They further concluded that after correcting for publication bias, and assuming that 40% of all studies are unpublished, this increased risk decreased from 24% to 15%. This conclusion has been challenged on the basis that the assumption that 40% of all studies are unpublished was "extreme". In 2006, Takagi et al. reanalyzed the data from this meta-analysis to account for publication bias and estimated that the relative risk of lung cancer among those exposed to secondhand smoke was 1.19, slightly lower than the original estimate. A 2000 meta-analysis found a relative risk of 1.48 for lung cancer among men exposed to secondhand smoke, and a relative risk of 1.16 among those exposed to it at work. Another meta-analysis confirmed the finding of an increased risk of lung cancer among women with spousal exposure to secondhand smoke the following year. It found a relative risk of lung cancer of 1.29 for women exposed to secondhand smoke from their spouses. A 2014 meta-analysis noted that "the association between exposure to secondhand smoke and lung cancer risk is well established."

A minority of epidemiologists have found it hard to understand how second-hand smoke, which is more diluted than actively inhaled smoke, could have an effect that is such a large fraction of the added risk of coronary heart disease among active smokers. One proposed explanation is that second-hand smoke is not simply a diluted version of "mainstream" smoke, but has a different composition with more toxic substances per gram of total particulate matter. Passive smoking appears to be capable of precipitating the acute manifestations of cardio-vascular diseases (atherothrombosis) and may also have a negative impact on the outcome of patients who suffer acute coronary syndromes.

In 2004, the International Agency for Research on Cancer (IARC) of the World Health Organization (WHO) reviewed all significant published evidence related to tobacco smoking and cancer. It concluded:

These meta-analyses show that there is a statistically significant and consistent association between lung cancer risk in spouses of smokers and exposure to second-hand tobacco smoke from the spouse who smokes. The excess risk is of the order of 20% for women and 30% for men and remains after controlling for some potential sources of bias and confounding.

Subsequent meta-analyses have confirmed these findings.

The National Asthma Council of Australia cites studies showing that second-hand smoke is probably the most important indoor pollutant, especially around young children:

- Smoking by either parent, particularly by the mother, increases the risk of asthma in children.

- The outlook for early childhood asthma is less favourable in smoking households.

- Children with asthma who are exposed to smoking in the home generally have more severe disease.

- Many adults with asthma identify ETS as a trigger for their symptoms.

- Doctor-diagnosed asthma is more common among non-smoking adults exposed to ETS than those not exposed. Among people with asthma, higher ETS exposure is associated with a greater risk of severe attacks.

In France, exposure to second-hand smoke has been estimated to cause between 3,000 and 5,000 premature deaths per year, with the larger figure cited by Prime Minister Dominique de Villepin during his announcement of a nationwide smoke-free law: "That makes more than 13 deaths a day. It is an unacceptable reality in our country in terms of public health."

There is good observational evidence that smoke-free legislation reduces the number of hospital admissions for heart disease.

Risk Level

The International Agency for Research on Cancer of the World Health Organization concluded in 2004 that there was sufficient evidence that second-hand smoke caused cancer in humans. Those who work in environments where smoke is not regulated are at higher risk. Workers particularly at risk of exposure include those in installation repair and maintenance, construction and extraction, and transportation.

The US Surgeon General, in his 2006 report, estimated that living or working in a place where smoking is permitted increases the non-smokers' risk of developing heart disease by 25–30% and lung cancer by 20–30%. Long term firsthand smoking increases the risk more than 1000%.

Biomarkers

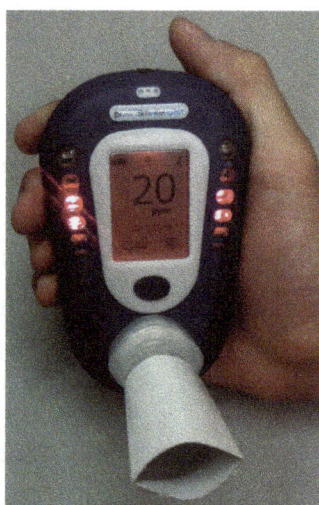

Breath CO monitor displaying carbon monoxide concentration of an exhaled breath sample (in ppm) with corresponding percent concentration of carboxyhemoglobin displayed below.

Environmental tobacco smoke can be evaluated either by directly measuring tobacco smoke pollutants found in the air or by using biomarkers, an indirect measure of exposure. Carbon monoxide monitored through breath, nicotine, cotinine, thiocyanates, and proteins are the most specific biological markers of tobacco smoke exposure. Biochemical tests are a much more reliable biomarker of second-hand smoke exposure than surveys. Certain groups of people are reluctant to disclose their smoking status and exposure to tobacco smoke, especially pregnant women and parents of young children. This is due to their smoking being socially unacceptable. Also, it may be difficult for individuals to recall their exposure to tobacco smoke.

A 2007 study in the *Addictive Behaviors* journal found a positive correlation between second-hand tobacco smoke exposure and concentrations of nicotine and/or biomarkers of nicotine in the body. Significant biological levels of nicotine from second-hand smoke exposure were equivalent to nicotine levels from active smoking and levels that are associated with behaviour changes due to nicotine consumption.

Cotinine

Cotinine, the metabolite of nicotine, is a biomarker of second-hand smoke exposure. Typically, cotinine is measured in the blood, saliva, and urine. Hair analysis has recently become a new, noninvasive measurement technique. Cotinine accumulates in hair during hair growth, which results in a measure of long-term, cumulative exposure to tobacco smoke. Urinary cotinine levels have been a reliable biomarker of tobacco exposure and have been used as a reference in many epidemiological studies. However, cotinine levels found in the urine only reflect exposure over the preceding 48 hours. Cotinine levels of the skin, such as the hair and nails, reflect tobacco exposure over the previous three months and are a more reliable biomarker.

Carbon Monoxide (CO)

Carbon monoxide monitored via breath is also a reliable biomarker of second-hand smoke exposure as well as tobacco use. With high sensitivity and specificity, it not only provides an accurate measure, but the test is also non-invasive, highly reproducible, and low in cost. Breath CO monitoring measures the concentration of CO in an exhalation in parts per million, and this can be directly correlated to the blood CO concentration (carboxyhemoglobin). Breath CO monitors can also be used by emergency services to identify patients who are suspected of having CO poisoning.

Pathophysiology

A 2004 study by the International Agency for Research on Cancer of the World Health Organization concluded that non-smokers are exposed to the same carcinogens as active smokers. Sidestream smoke contains more than 4,000 chemicals, including 69 known carcinogens. Of special concern are polynuclear aromatic hydrocarbons, tobacco-specific N-nitrosamines, and aromatic amines, such as 4-aminobiphenyl, all known to be highly carcinogenic. Mainstream smoke, sidestream smoke, and second-hand smoke contain largely the same components, however the concentration varies depending on type of smoke. Several well-established carcinogens have been shown by the tobacco companies' own research to be present at higher concentrations in sidestream smoke than in mainstream smoke.

Second-hand smoke has been shown to produce more particulate-matter (PM) pollution than an idling low-emission diesel engine. In an experiment conducted by the Italian National Cancer Institute, three cigarettes were left smoldering, one after the other, in a 60 m³ garage with a limited air exchange. The cigarettes produced PM pollution exceeding outdoor limits, as well as PM concentrations up to 10-fold that of the idling engine.

Second-hand tobacco smoke exposure has immediate and substantial effects on blood and blood vessels in a way that increases the risk of a heart attack, particularly in people already at risk. Exposure to tobacco smoke for 30 minutes significantly reduces coronary flow velocity reserve in healthy nonsmokers. Second-hand smoke is also associated with impaired vasodilation among adult nonsmokers. Second-hand smoke exposure also affects platelet function, vascular endothelium, and myocardial exercise tolerance at levels commonly found in the workplace.

Pulmonary emphysema can be induced in rats through acute exposure to sidestream tobacco smoke (30 cigarettes per day) over a period of 45 days. Degranulation of mast cells contributing to lung damage has also been observed.

The term "third-hand smoke" was recently coined to identify the residual tobacco smoke contamination that remains after the cigarette is extinguished and second-hand smoke has cleared from the air. Preliminary research suggests that by-products of third-hand smoke may pose a health risk, though the magnitude of risk, if any, remains unknown. In October 2011, it was reported that Christus St. Frances Cabrini Hospital in Alexandria, Louisiana would seek to eliminate third-hand smoke beginning in July 2012, and that employees whose clothing smelled of smoke would not be allowed to work. This prohibition was enacted because third-hand smoke poses a special danger for the developing brains of infants and small children.

In 2008, there were more than 161,000 deaths attributed to lung cancer in the United States. Of these deaths, an estimated 10% to 15% were caused by factors other than first-hand smoking; equivalent to 16,000 to 24,000 deaths annually. Slightly more than half of the lung cancer deaths caused by factors other than first-hand smoking were found in nonsmokers. Lung cancer in non-smokers may well be considered one of the most common cancer mortalities in the United States. Clinical epidemiology of lung cancer has linked the primary factors closely tied to lung cancer in non-smokers as exposure to second-hand tobacco smoke, carcinogens including radon, and other indoor air pollutants.

Opinion of Public Health Authorities

There is widespread scientific consensus that exposure to second-hand smoke is harmful. The link between passive smoking and health risks is accepted by every major medical and scientific organisation, including:

- World Health Organization

- U.S. National Institutes of Health

- Centers for Disease Control

- United States Surgeon General

- U.S. National Cancer Institute

- United States Environmental Protection Agency

- California Environmental Protection Agency

- American Heart Association, American Lung Association, and American Cancer Society

- American Medical Association

- American Academy of Pediatrics

- Australian National Health and Medical Research Council

- United Kingdom Scientific Committee on Tobacco and Health

Public Opinion

Recent major surveys conducted by the U.S. National Cancer Institute and Centers for Disease Control have found widespread public awareness that second-hand smoke is harmful. In both 1992 and 2000 surveys, more than 80% of respondents agreed with the statement that second-hand smoke was harmful. A 2001 study found that 95% of adults agreed that second-hand smoke was harmful to children, and 96% considered tobacco-industry claims that second-hand smoke was not harmful to be untruthful.

A 2007 Gallup poll found that 56% of respondents felt that second-hand smoke was "very harmful", a number that has held relatively steady since 1997. Another 29% believe that second-hand smoke is "somewhat harmful"; 10% answered "not too harmful", while 5% said "not at all harmful".

Controversy Over Harm

As part of its attempt to prevent or delay tighter regulation of smoking, the tobacco industry funded a number of scientific studies and, where the results cast doubt on the risks associated with second-hand smoke, sought wide publicity for those results. The industry also funded libertarian and conservative think tanks, such as the Cato Institute in the United States and the Institute of Public Affairs in Australia which criticised both scientific research on passive smoking and policy proposals to restrict smoking. *New Scientist* and the *European Journal of Public Health* have identified these industry-wide coordinated activities as one of the earliest expressions of corporate denialism. Further, they state that the disinformation spread by the tobacco industry has created a *tobacco denialism* movement, sharing many characteristics of other forms of denialism, such as HIV-AIDS denialism.

Industry-Funded Studies and Critiques

Enstrom and Kabat

A 2003 study by James Enstrom and Geoffrey Kabat, published in the *British Medical*

Journal, argued that the harms of passive smoking had been overstated. Their analysis reported no statistically significant relationship between passive smoking and lung cancer, coronary heart disease (CHD), or chronic obstructive pulmonary disease, though the accompanying editorial noted that "they may overemphasise the negative nature of their findings." This paper was widely promoted by the tobacco industry as evidence that the harms of passive smoking were unproven. The American Cancer Society (ACS), whose database Enstrom and Kabat used to compile their data, criticized the paper as "neither reliable nor independent", stating that scientists at the ACS had repeatedly pointed out serious flaws in Enstrom and Kabat's methodology prior to publication. Notably, the study had failed to identify a comparison group of "unexposed" persons.

Enstrom's ties to the tobacco industry also drew scrutiny; in a 1997 letter to Philip Morris, Enstrom requested a "substantial research commitment... in order for me to effectively compete against the large mountain of epidemiologic data and opinions that already exist regarding the health effects of ETS and active smoking." In a US racketeering lawsuit against tobacco companies, the Enstrom and Kabat paper was cited by the US District Court as "a prime example of how nine tobacco companies engaged in criminal racketeering and fraud to hide the dangers of tobacco smoke." The Court found that the study had been funded and managed by the Center for Indoor Air Research, a tobacco industry front group tasked with "offsetting" damaging studies on passive smoking, as well as by Philip Morris who stated that Enstrom's work was "clearly litigation-oriented." A 2005 paper in *Tobacco Control* argued that the disclosure section in the Enstrom and Kabat BMJ paper, although it met the journal's requirements, "does not reveal the full extent of the relationship the authors had with the tobacco industry."

In 2006, Enstrom and Kabat published a meta-analysis of studies regarding passive smoking and coronary heart disease in which they reported a very weak association between passive smoking and heart disease mortality. They concluded that exposure to second-hand smoke increased the risk of death from CHD by only 5%, although this analysis has been criticized for including two previous industry-funded studies that suffered from widespread exposure misclassification.

Gori

Gio Batta Gori, a tobacco industry spokesman and consultant and an expert on risk utility and scientific research, wrote in the libertarian Cato Institute's magazine *Regulation* that "...of the 75 published studies of ETS and lung cancer, some 70 percent did not report statistically significant differences of risk and are moot. Roughly 17 percent claim an increased risk and 13 percent imply a reduction of risk."

Milloy

Steven Milloy, the "junk science" commentator for Fox News and a former Philip Morris consultant, claimed that "of the 19 studies" on passive smoking "only 8— slightly more than 42 percent— reported statistically significant increases in heart disease incidence.."

Another component of criticism cited by Milloy focused on relative risk and epidemiological practices in studies of passive smoking. Milloy, who has a master's degree from the Johns Hopkins School of Hygiene and Public Health, argued that studies yielding relative risks of less than 2 were meaningless junk science. This approach to epidemiological analysis was criticized in the *American Journal of Public Health*:

A major component of the industry attack was the mounting of a campaign to establish a "bar" for "sound science" that could not be fully met by most individual investigations, leaving studies that did not meet the criteria to be dismissed as "junk science."

The tobacco industry and affiliated scientists also put forward a set of "Good Epidemiology Practices" which would have the practical effect of obscuring the link between secondhand smoke and lung cancer; the privately stated goal of these standards was to "impede adverse legislation". However, this effort was largely abandoned when it became clear that no independent epidemiological organization would agree to the standards proposed by Philip Morris et al.

Levois and Layard

In 1995, Levois and Layard, both tobacco industry consultants, published two analyses in the journal *Regulatory Toxicology and Pharmacology* regarding the association between spousal exposure to second-hand smoke and heart disease. Both of these papers reported no association between second-hand smoke and heart disease. These analyses have been criticized for failing to distinguish between current and former smokers, despite the fact that former smokers, unlike current ones, are not at a significantly increased risk of heart disease.

World Health Organization Controversy

A 1998 study by the International Agency for Research on Cancer (IARC) on environmental tobacco smoke (ETS) found "weak evidence of a dose-response relationship between risk of lung cancer and exposure to spousal and workplace ETS."

In March 1998, before the study was published, reports appeared in the media alleging that the IARC and the World Health Organization (WHO) were suppressing information. The reports, appearing in the British *Sunday Telegraph* and *The Economist*, among other sources, alleged that the WHO withheld from publication of its own report that supposedly failed to prove an association between passive smoking and a number of other diseases (lung cancer in particular).

In response, the WHO issued a press release stating that the results of the study had been "completely misrepresented" in the popular press and were in fact very much in line with similar studies demonstrating the harms of passive smoking. The study was published in the *Journal of the National Cancer Institute* in October of the same year, and concluded the authors found "no association between childhood exposure to ETS and lung cancer risk" but "did find weak evidence of a dose–response relationship between risk of lung cancer and exposure to spousal and workplace ETS." An accompanying editorial summarized:

When all the evidence, including the important new data reported in this issue of the Journal, is assessed, the inescapable scientific conclusion is that ETS is a low-level lung carcinogen.

With the release of formerly classified tobacco industry documents through the Tobacco Master Settlement Agreement, it was found (by Elisa Ong and Stanton Glantz) that the controversy over the WHO's alleged suppression of data had been engineered by Philip Morris, British American Tobacco, and other tobacco companies in an effort to discredit scientific findings which would harm their business interests. A WHO inquiry, conducted after the release of the tobacco-industry documents, found that this controversy was generated by the tobacco industry as part of its larger campaign to cut the WHO's budget, distort the results of scientific studies on passive smoking, and discredit the WHO as an institution. This campaign was carried out using a network of ostensibly independent front organizations and international and scientific experts with hidden financial ties to the industry.

EPA Lawsuit

In 1993, the United States Environmental Protection Agency (EPA) issued a report estimating that 3,000 lung cancer related deaths in the United States were caused by passive smoking annually.

Philip Morris, R.J. Reynolds Tobacco Company, and groups representing growers, distributors and marketers of tobacco took legal action, claiming that the EPA had manipulated this study and ignored accepted scientific and statistical practices.

The United States District Court for the Middle District of North Carolina ruled in favor of the tobacco industry in 1998, finding that the EPA had failed to follow proper scientific and epidemiologic practices and had "cherry picked" evidence to support conclusions which they had committed to in advance. The court stated in part, "EPA publicly committed to a conclusion before research had begun...adjusted established procedure and scientific norms to validate the Agency's public conclusion... In conducting the ETS Risk Assessment, disregarded information and made findings on selective information; did not disseminate significant epidemiologic information; deviated from its Risk Assessment Guidelines; failed to disclose important findings and reasoning..."

In 2002, the EPA successfully appealed this decision to the United States Court of Appeals for the Fourth Circuit. The EPA's appeal was upheld on the preliminary grounds that their report had no regulatory weight, and the earlier finding was vacated.

In 1998, the U.S. Department of Health and Human Services, through the publication by its National Toxicology Program of the 9th Report on Carcinogens, listed environmental tobacco smoke among the known carcinogens, observing of the EPA assessment that "The individual studies were carefully summarized and evaluated."

Tobacco-industry Funding of Research

The tobacco industry's role in funding scientific research on second-hand smoke has been controversial. A review of published studies found that tobacco-industry affiliation was strongly correlated with findings exonerating second-hand smoke; researchers affiliated with the tobacco industry were 88 times more likely than independent researchers to conclude that second-hand smoke was not harmful. In a specific example which came to light with the release of tobacco-industry documents, Philip Morris executives successfully encouraged an author to revise his industry-funded review article to downplay the role of second-hand smoke in sudden infant death syndrome. The 2006 U.S. Surgeon General's report criticized the tobacco industry's role in the scientific debate:

The industry has funded or carried out research that has been judged to be biased, supported scientists to generate letters to editors that criticized research publications, attempted to undermine the findings of key studies, assisted in establishing a scientific society with a journal, and attempted to sustain controversy even as the scientific community reached consensus.

This strategy was outlined at an international meeting of tobacco companies in 1988, at which Philip Morris proposed to set up a team of scientists, organized by company lawyers, to "carry out work on ETS to keep the controversy alive." All scientific research was subject to oversight and "filtering" by tobacco-industry lawyers:

Philip Morris then expect the group of scientists to operate within the confines of decisions taken by PM scientists to determine the general direction of research, which apparently would then be 'filtered' by lawyers to eliminate areas of sensitivity.

Philip Morris reported that it was putting "...vast amounts of funding into these projects... in attempting to coordinate and pay so many scientists on an international basis to keep the ETS controversy alive."

Tobacco Industry Response

Measures to tackle second-hand smoke pose a serious economic threat to the tobacco industry, having broadened the definition of smoking beyond a personal habit to something with a social impact. In a confidential 1978 report, the tobacco industry described

increasing public concerns about second-hand smoke as "the most dangerous development to the viability of the tobacco industry that has yet occurred." In *United States of
America v. Philip Morris et al.*, the District Court for the District of Columbia found
that the tobacco industry "... recognized from the mid-1970s forward that the health
effects of passive smoking posed a profound threat to industry viability and cigarette
profits," and that the industry responded with "efforts to undermine and discredit the
scientific consensus that ETS causes disease."

Accordingly, the tobacco industry have developed several strategies to minimise the
impact on their business:

- The industry has sought to position the second-hand smoke debate as essentially concerned with civil liberties and smokers' rights rather than with health, by
  funding groups such as FOREST.

- Funding bias in research; in all reviews of the effects of second-hand smoke
  on health published between 1980 and 1995, the only factor associated with
  concluding that second-hand smoke is not harmful was whether an author was
  affiliated with the tobacco industry. However, not all studies that failed to find
  evidence of harm were by industry-affiliated authors.

- Delaying and discrediting legitimate research (see for an example of how the
  industry attempted to discredit Takeshi Hirayama's landmark study, and for an
  example of how it attempted to delay and discredit a major Australian report on
  passive smoking)

- Promoting "good epidemiology" and attacking so-called junk science (a term
  popularised by industry lobbyist Steven Milloy): attacking the methodology behind research showing health risks as flawed and attempting to promote sound
  science. Ong & Glantz (2001) cite an internal Phillip Morris memo giving evidence of this as company policy.

- Creation of outlets for favourable research. In 1989, the tobacco industry
  established the International Society of the Built Environment, which published the peer-reviewed journal *Indoor and Built Environment*. This journal did not require conflict-of-interest disclosures from its authors. With
  documents made available through the Master Settlement, it was found that
  the executive board of the society and the editorial board of the journal were
  dominated by paid tobacco-industry consultants. The journal published a
  large amount of material on passive smoking, much of which was "indus-
  try-positive".

Citing the tobacco industry's production of biased research and efforts to undermine
scientific findings, the 2006 U.S. Surgeon General's report concluded that the industry
had "attempted to sustain controversy even as the scientific community reached con-

sensus... industry documents indicate that the tobacco industry has engaged in widespread activities... that have gone beyond the bounds of accepted scientific practice." The U.S. District Court, in *U.S.A. v. Philip Morris et al.*, found that "...despite their internal acknowledgment of the hazards of secondhand smoke, Defendants have fraudulently denied that ETS causes disease."

Position of Major Tobacco Companies

The positions of major tobacco companies on the issue of second-hand smoke is somewhat varied. In general, tobacco companies have continued to focus on questioning the methodology of studies showing that second-hand smoke is harmful. Some (such as British American Tobacco and Philip Morris) acknowledge the medical consensus that second-hand smoke carries health risks, while others continue to assert that the evidence is inconclusive. Several tobacco companies advocate the creation of smoke-free areas within public buildings as an alternative to comprehensive smoke-free laws.

US racketeering Lawsuit Against Tobacco Companies

On September 22, 1999, the U.S. Department of Justice filed a racketeering lawsuit against Philip Morris and other major cigarette manufacturers. Almost 7 years later, on August 17, 2006 U.S. District Court Judge Gladys Kessler found that the Government had proven its case and that the tobacco company defendants had violated the Racketeer Influenced Corrupt Organizations Act (RICO). In particular, Judge Kessler found that PM and other tobacco companies had:

- conspired to minimize, distort and confuse the public about the health hazards of smoking;

- publicly denied, while internally acknowledging, that second-hand tobacco smoke is harmful to nonsmokers, and

- destroyed documents relevant to litigation.

The ruling found that tobacco companies undertook joint efforts to undermine and discredit the scientific consensus that second-hand smoke causes disease, notably by controlling research findings via paid consultants. The ruling also concluded that tobacco companies were fraudulently continuing to deny the health effects of ETS exposure.

On May 22, 2009, a three-judge panel of the U.S. Court of Appeals for the District of Columbia Circuit unanimously upheld the lower court's 2006 ruling.

Smoke-free Laws

As a consequence of the health risks associated with second-hand smoke, smoke-free regulations in indoor public places, including restaurants, cafés, and nightclubs have been introduced in a number of jurisdictions, at national or local level, as well as some outdoor open areas. Ireland was the first country in the world to institute a comprehensive national smoke-free law on smoking in all indoor workplaces on 29 March 2004. Since then, many others have followed suit. The countries which have ratified the WHO Framework Convention on Tobacco Control (FCTC) have a legal obligation to implement *effective* legislation "for protection from exposure to tobacco smoke in indoor workplaces, public transport, indoor public places and, as appropriate, other public places." (Article 8 of the FCTC) The parties to the FCTC have further adopted *Guidelines on the Protection from Exposure to Second hand Smoke* which state that "effective measures to provide protection from exposure to tobacco smoke ... require the total elimination of smoking and tobacco smoke in a particular space or environment in order to create a 100% smoke-free environment."

Opinion polls have shown considerable support for smoke-free laws. In June 2007, a survey of 15 countries found 80% approval for smoke-free laws. A survey in France, reputedly a nation of smokers, showed 70% support.

Effects

Smoking bans by governments result in decreased harm from second hand smoke including decrease cardiovascular disease. In the first 18 months after the town of Pueblo, Colorado enacted a smoke-free law in 2003, hospital admissions for heart attacks dropped 27%. Admissions in neighbouring towns without smoke-free laws showed no change, and the decline in heart attacks in Pueblo was attributed to the resulting reduction in second-hand smoke exposure. A 2004 smoking ban instituted in Massachusetts workplaces decreased workers' secondhand smoke exposure from 8% of workers in 2003 to 5.4% of workers in 2010. A 2016 review also found benefits of decrease exposure to smoke from specific location policies.

In 2001, a systematic review for the Guide to Community Preventative Services acknowledged strong evidence of the effectiveness of smoke-free policies and restrictions in reducing expose to second-hand smoke. A follow up to this review, identified the evidence on which the effectiveness of smoking bans reduced the prevalence of tobacco use. Articles published until 2005, were examined to further support this evidence. The examined studies provided sufficient evidence that smoke-free policies reduce tobacco use among workers when implemented in worksites or by communities.

While a number of studies funded by the tobacco industry have claimed a negative economic impact from smoke-free laws, no independently funded research has shown any such impact. A 2003 review reported that independently funded, methodologically sound research consistently found either no economic impact or a positive impact from smoke-free laws.

Air nicotine levels were measured in Guatemalan bars and restaurants before and after an implemented smoke-free law in 2009. Nicotine concentrations significantly decreased in both the bars and restaurants measured. Also, the employees support for a smoke-free workplace substantially increased in the post-implementation survey compared to pre-implementation survey. The result of this smoke-free law provides a considerably more healthy work environment for the staff.

Public Opinion

Recent surveys taken by the Society for Research on Nicotine and Tobacco demonstrates supportive attitudes of the public, towards smoke-free policies in outdoor areas. A vast majority of the public supports restricting smoking in various outdoor settings. The respondents reasons for supporting the polices were for varying reasons such as, litter control, establishing positive smoke-free role models for youth, reducing youth opportunities to smoke, and avoiding exposure to secondhand smoke.

Alternative Forms

Alternatives to smoke-free laws have also been proposed as a means of harm reduction, particularly in bars and restaurants. For example, critics of smoke-free laws cite studies suggesting ventilation as a means of reducing tobacco smoke pollutants and improving air quality. Ventilation has also been heavily promoted by the tobacco industry as an alternative to outright bans, via a network of ostensibly independent experts with often undisclosed ties to the industry. However, not all critics have connections to the industry.

The American Society of Heating, Refrigerating and Air-Conditioning Engineers (ASHRAE) officially concluded in 2005 that while completely isolated smoking rooms do eliminate the risk to nearby non-smoking areas, smoking bans are the only means of completely eliminating health risks associated with indoor exposure. They further concluded that no system of dilution or cleaning was effective at eliminating risk. The U.S. Surgeon General and the European Commission Joint Research Centre have reached similar conclusions. The implementation guidelines for the WHO Framework Convention on Tobacco Control states that engineering approaches, such as ventilation, are ineffective and do not protect against second-hand smoke exposure. However, this does *not* necessarily mean that such measures are useless in reducing harm, only that they fall short of the goal of reducing exposure completely to zero.

Others have suggested a system of tradable smoking pollution permits, similar to the cap-and-trade pollution permits systems used by the Environmental Protection Agency in recent decades to curb other types of pollution. This would guarantee that a portion of bars/restaurants in a jurisdiction will be smoke-free, while leaving the decision to the market.

In Animals

Multiple studies have been conducted to determine the carcinogenicity of environmental tobacco smoke to animals. These studies typically fall under the categories of simulated environmental tobacco smoke, administering condensates of sidestream smoke, or observational studies of cancer among pets.

To simulate environmental tobacco smoke, scientists expose animals to sidestream smoke, that which emanates from the cigarette's burning cone and through its paper, or a combination of mainstream and sidestream smoke. The IARC monographs conclude that mice with prolonged exposure to simulated environmental tobacco smoke, that is 6hrs a day, 5 days a week, for five months with a subsequent 4 month interval before dissection, will have significantly higher incidence and multiplicity of lung tumors than with control groups.

The IARC monographs concluded that sidestream smoke condensates had a significantly higher carcinogenic effect on mice than did mainstream smoke condensates.

Observational Studies

Second-hand smoke is popularly recognised as a risk factor for cancer in pets. A study conducted by the Tufts University School of Veterinary Medicine and the University of Massachusetts Amherst linked the occurrence of feline oral cancer to exposure to environmental tobacco smoke through an overexpression of the p53 gene. Another study conducted at the same universities concluded that cats living with a smoker were more likely to get feline lymphoma; the risk increased with the duration of exposure to secondhand smoke and the number of smokers in the household. A study by Colorado State University researchers, looking at cases of canine lung cancer, was generally inconclusive, though the authors reported a weak relation for lung cancer in dogs exposed to environmental tobacco smoke. The number of smokers within the home, the number of packs smoked in the home per day, and the amount of time that the dog spent within the home had no effect on the dog's risk for lung cancer.

Terminology

As of 2003, "secondhand smoke" was the term most used to refer to other people's smoke in the English-language media. Other terms used include "environmental tobacco smoke", while "involuntary smoking" and "passive smoking" are used to refer to exposure to secondhand smoke. The term "environmental tobacco smoke" can be traced back to a 1974 industry-sponsored meeting held in Bermuda, while the term "passive smoking" was first used in the title of a scientific paper in 1970. The Surgeon General prefers to use the phrase "secondhand smoke" rather than "environmental tobacco smoke", stating that "The descriptor "secondhand" captures the involuntary nature of the exposure, while "environmental" does not." Most researchers consider the term "passive smoking" to be synonymous with "secondhand smoke". In contrast, a 2011 commentary in *Environmental Health Perspectives* argued that research into "thirdhand smoke" renders it inappropriate to refer to passive smoking with the term "secondhand smoke", which the authors stated constitutes a pars pro toto.

Mold Indoor Growth, Assessment, and Remediation

Mold (American English) or mould (British English) is part of the natural environment. Outdoors, molds play a part in nature by breaking down dead organic matter such as fallen leaves and dead trees; indoors, mold growth should be avoided. Molds reproduce by means of tiny spores. The spores are invisible to the naked eye and float through the air. Mold may begin growing indoors when spores land on moist surfaces. There are many types of mold, but all require moisture for growth.

Health Effects

Molds are ubiquitous, and mold spores are a common component of household and workplace dust. In large amounts they can be a health hazard to humans, potentially causing allergic reactions and respiratory problems.

Some molds produce mycotoxins that can pose serious health risks to humans and animals. "Toxic mold" refers to molds which produce mycotoxins, such as *Stachybotrys chartarum*. Exposure to high levels of mycotoxins can lead to neurological problems and death. Prolonged exposure (for example, daily exposure) can be particularly harmful.

Symptoms

Symptoms of mold exposure may include:

- Nasal and sinus congestion; runny nose

- Eye irritation; itchy, red, watery eyes

- Respiratory problems, such as wheezing and difficulty breathing, chest tightness

- Cough

- Throat irritation

- Skin irritation, such as a rash

- Headache

- Persistent sneezing

Asthma

Infants may develop respiratory symptoms as a result of exposure to *Penicillium*, a fungal genus. Signs of mold-related respiratory problems in an infant include a persistent cough or wheeze. Increased exposure increases the probability of developing respiratory symptoms during the first year of life. Studies have indicated a correlation between the probability of developing asthma and exposure to *Penicillium*.

Mold exposure has a variety of health effects, and sensitivity to mold varies. Exposure to mold may cause throat irritation, nasal stuffiness, eye irritation, cough and wheezing and skin irritation in some cases. Exposure to mold may heighten sensitivity, depending on the time and nature of exposure. People with chronic lung diseases are at higher risk for mold allergies, and will experience more severe reactions when exposed to mold. Damp indoor environments correlate with upper-respiratory-tract symptoms, such as coughing and wheezing in people with asthma.

Causes and Growing Conditions

Molds are found everywhere, and can grow on almost any substance when moisture is present. They reproduce by spores, which are carried by air currents. When spores land on a moist surface suitable for life, they begin to grow. Mold is normally found indoors at levels which do not affect most healthy individuals.

Because common building materials are capable of sustaining mold growth and mold spores are ubiquitous, mold growth in an indoor environment is typically related to water or moisture and may be caused by incomplete drying of flooring materials (such as concrete). Flooding, leaky roofs, building-maintenance or indoor-plumbing problems can lead to interior mold growth. Water vapor commonly condenses on surfaces cooler than the moisture-laden air, enabling mold to flourish. This moisture vapor passes through walls and ceilings, typically condensing during the winter in climates with a long heating season. Floors over crawl spaces and basements, without vapor barriers or with dirt floors, are mold-prone. The "doormat test" detects moisture from concrete slabs without a sub-slab vapor barrier. Some materials, such a polished concrete, do not support mold growth.

Although this home experienced minor exterior damage from Hurricane Katrina, small leaks and inadequate airflow permitted mold infestation.

Significant mold growth requires moisture and food sources and a substrate capable of sustaining growth. Common building materials, such as plywood, drywall, furring strips, carpets, and carpet padding provide food for mold. In carpet, invisible dust and cellulose are food sources. After water damage to a building, mold grows in walls and then becomes dormant until subsequent high humidity; suitable conditions reactivate mold. Mycotoxin levels are higher in buildings which have had a water incident.

Hidden Mold

Mold is detectable by smell and signs of water damage on walls or ceiling, and can grow in places invisible to the human eye. It may be found behind wallpaper or paneling, on the inside of ceiling tiles, the back of drywall, or the underside of carpets or carpet padding. Piping in walls may also be a source of mold, since they may leak (causing moisture and condensation).

Spores need three things to grow into mold:

- Nutrients: Cellulose (the cell wall of green plants) is a common food for indoor spores.

- Moisture: To begin the decaying process caused by mold

- Time: Mold growth begins from 24 hours to 10 days after the provision of growing conditions.

Mold colonies can grow inside buildings, and the chief hazard is the inhalation of mycotoxins. After a flood or major leak, mycotoxin levels are higher even after a building has dried out.

Food sources for mold in buildings include cellulose-based materials such as wood, cardboard and the paper facing on drywall and organic matter such as soap, fabrics and dust-containing skin cells. If a house has mold, the moisture may originate in the basement or crawl space, a leaking roof or a leak in plumbing pipes. Insufficient ventilation may accelerate moisture buildup. Visible mold colonies may form where ventilation is poorest and on perimeter walls (because they are nearest the dew point).

If there are mold problems in a house only during certain times of the year, the house is probably too airtight or too drafty. Mold problems occur in airtight homes more frequently in the warmer months (when humidity is high inside the house, and moisture is trapped), and occur in drafty homes more frequently in the colder months (when warm air escapes from the living area and condenses). If a house is artificially humidified during the winter, this can create conditions favorable to mold. Moving air may prevent mold from growing, since it has the same desiccating effect as low humidity. Molds grow best in warm temperatures, 77 to 86 °F (25 to 30 °C), although growth may occur between 32 and 95 °F (0 and 35 °C).

Removing one of the three requirements for mold reduces (or eliminates) new mold growth:

- Moisture

- Food for the mold spores (for example, dust or dander)

- Warmth; mold generally does not grow in cold environments.

HVAC systems can produce all three requirements for mold growth. The air conditioning system creates a difference in temperature, encouraging condensation. The high rate of dusty air movement through an HVAC system may furnish ample food for mold. Since the air-conditioning system is not always running, warm conditions are the final component for mold growth.

Assessment

The first step in assessment is to non-intrusively determine if mold is present by visually examining the premises; visible mold helps determine the level of remediation necessary. If mold is actively growing and visibly confirmed, sampling for its specific species is unnecessary.

Intrusive observation is sometimes needed to assess the mold level. This includes moving furniture, lifting (or removing) carpets, checking behind wallpaper or paneling, checking ventilation ductwork and exposing wall cavities. Detailed visual inspection and the recognition of moldy odors should be used to find problems. Efforts should focus on areas where there are signs of liquid moisture or water vapor (humidity), or where moisture problems are suspected.

Sampling

The United States Environmental Protection Agency (EPA) does not generally recommend sampling unless an occupant of the space has symptoms. Sampling should be performed by a trained professional with specific experience in mold-sampling protocols, sampling methods and the interpretation of findings. It should be done only to make a particular determination, such as airborne spore concentration or identifying a particular species. Before sampling, a subsequent course of action should be determined.

In the U.S., sampling and analysis should follow the recommendations of the Occupational Safety and Health Administration (OSHA), National Institute for Occupational Safety and Health (NIOSH), the EPA and the American Industrial Hygiene Association (AIHA).

Types of samples include:

- Air: The most common form of sampling to assess mold levels. Indoor and outdoor air are sampled, and their mold-spore levels compared. Air sampling often identifies hidden mold.

- Surface: Measures the number of mold spores deposited on indoor surfaces, collected on tape or in dust

- Bulk: Removal of material from the contaminated area to identify and quantify the mold in the sample

- Swab: Something akin to a cotton swab is rubbed across the area being sampled, often a measured area, and subsequently sent to the mold testing laboratory. Final results will indicate mold levels and species located in suspect area.

Multiple types of sampling are recommended by the AIHA, since each has limitations; for example, air samples will not identify a hidden mold source and a tape sample cannot determine the level of contamination in the air.

Remediation

Mold remediation

The first step in solving an indoor mold problem is to remove the moisture source; new mold will begin to grow on moist, porous surfaces within 24 to 48 hours. There are a number of ways to prevent mold growth. Some cleaning companies specialize in fabric restoration, removing mold (and mold spores) from clothing to eliminate odor and prevent further damage to garments.

The effective way to clean mold is to use detergent solutions which physically remove mold. Many commercially available detergents marketed for mold cleanup include an EPA-approved antifungal agent.

Significant mold growth may require professional mold remediation to remove the affected building materials and eradicate the source of excess moisture. In extreme cases of mold growth in buildings, it may be more cost-effective to condemn the building than to reduce mold to safe levels.

The goals of remediation are to remove (or clean) contaminated materials, preventing fungi (and fungi-contaminated dust) from entering an occupied (or non-contaminated) area while protecting workers performing the abatement.

Cleanup and Removal Methods

The purpose of cleanup is to eliminate mold and remove contaminated materials. Killing mold with a biocide is insufficient, since chemicals and proteins causing reactions in humans remain in dead mold. The following methods are used:

- Evaluation: Before remediation, the area is assessed to ensure safety, clean up the entire moldy area, and properly approach the mold. The EPA provides the following instructions:

- HVAC cleaning: Should be done by a trained professional.

- Protective clothing: Includes a half- or full-face respirator mask. Goggles with a half-face respirator mask prevent mold spores from reaching the mucous membranes of the eyes. Disposable hazmat coveralls are available to keep out particles down to one micrometer, and protective suits keep mold spores from entering skin cuts. Gloves are made of rubber, nitrile, polyurethane, or neoprene.

- Dry brushing or agitation device: Wire brushing or sanding is used when microbial growth can be seen on solid wood surfaces such as framing or underlayment (the subfloor).

- Dry-ice blasting: Removes mold from wood and cement; however, this process may spray mold and its bi-products into surrounding air.

- Wet vacuum: Wet vacuuming is used on wet materials, and this method is one of those approved by the EPA.

- Damp wipe: Removal of mold from non-porous surfaces by wiping or scrubbing with water and a detergent and drying quickly

- HEPA (high-efficiency particulate air) vacuum: Used in remediation areas after materials have been dried and contaminated materials removed; collected debris and dust is stored to prevent debris release.

- Debris disposal: Sealed in the remediation area, debris is usually discarded with ordinary construction waste.

Equipment

Equipment used in mold remediation includes:

- Moisture meter: Measures drying of damaged materials

- Humidity gauge: Often paired with a thermometer

- Borescope: Camera at the end of a flexible snake, illuminating potential mold problems inside walls, ceilings and crawl spaces

- Digital camera: Documents findings during assessment

- Personal protective equipment (PPE): Respirators, gloves, impervious suit, and eye protection

- Thermographic camera : Infrared thermal-imaging cameras identify secondary moisture sources.

Protection Levels

During mold remediation in the U.S., the level of contamination dictates the protection level for remediation workers. Contamination levels have been enumerated as I, II, III, and IV:

- *Level I*: Small, isolated areas (10 square feet (0.93 m²) or less); remediation may be conducted by trained building staff.

- *Level II*: Mid-sized, isolated areas (10–30 square feet (0.93–2.79 m²)); may also be remediated by trained, protected building staff.

- *Level III*: Large, isolated areas (30–100 square feet (2.8–9.3 m²)): Professionals experienced in microbial investigations or mold remediation should be consulted, and personnel should be trained in the handling of hazardous materials and equipped with respiratory protection, gloves and eye protection.

- *Level IV*: Extensive contamination (more than 100 square feet (9.3 m²)); requires trained, equipped professionals

After remediation, the premises should be reevaluated to ensure success.

Residential Mold Prevention and Control

According to the EPA, residential mold may be prevented and controlled in the following ways:

- Cleaning and repairing roof gutters, to prevent moisture seepage into the home

- Keeping air-conditioning drip pans clean and drainage lines clear

- Monitoring indoor humidity

- Drying areas of moisture or condensation and removing their sources

- Treating exposed structural wood or wood framing with an EPA-approved fungicidal encapsulation coating after pre-cleaning (particularly homes with a crawl space, unfinished basement or a poorly-ventilated attic)

HVAC

Rooftop HVAC unit with view of fresh air intake vent.

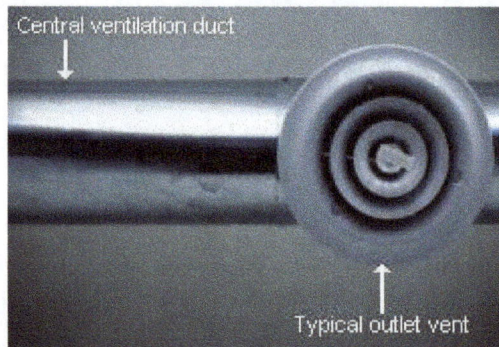

Ventilation duct with outlet vent. These are installed throughout a building to move air in or out of a room.

Heating, ventilation and air conditioning (HVAC) is the technology of indoor and vehicular environmental comfort. Its goal is to provide thermal comfort and acceptable indoor air quality. HVAC system design is a subdiscipline of mechanical engineering, based on the principles of thermodynamics, fluid mechanics, and heat transfer. Refrigeration is sometimes added to the field's abbreviation as HVAC&R or HVACR, or ventilating is dropped as in HACR (such as the designation of HACR-rated circuit breakers).

HVAC is an important part of residential structures such as single family homes, apartment buildings, hotels and senior living facilities, medium to large industrial and office buildings such as skyscrapers and hospitals, onboard vessels, and in marine environments, where safe and healthy building conditions are regulated with respect to temperature and humidity, using fresh air from outdoors.

Ventilating or ventilation (the *V* in HVAC) is the process of exchanging or replacing air in any space to provide high indoor air quality which involves temperature control, oxygen replenishment, and removal of moisture, odors, smoke, heat, dust, airborne bacteria, carbon dioxide, and other gases. Ventilation removes unpleasant smells and excessive moisture, introduces outside air, keeps interior building air circulating, and prevents stagnation of the interior air.

Ventilation includes both the exchange of air to the outside as well as circulation of air within the building. It is one of the most important factors for maintaining acceptable indoor air quality in buildings. Methods for ventilating a building may be divided into *mechanical/forced* and *natural* types.

Overview

The three central functions of heating, ventilation, and air conditioning are interrelated, especially with the need to provide thermal comfort and acceptable indoor air quality within reasonable installation, operation, and maintenance costs. HVAC systems can provide ventilation, reduce air infiltration, and maintain pressure relationships between spaces. The means of air delivery and removal from spaces is known as room air distribution.

Individual Systems

In modern buildings the design, installation, and control systems of these functions are integrated into one or more HVAC systems. For very small buildings, contractors normally estimate the capacity, engineer, and select HVAC systems and equipment. For larger buildings, building service designers, mechanical engineers, or building services engineers analyze, design, and specify the HVAC systems. Specialty mechanical contractors then fabricate and commission the systems. Building permits and code-compliance inspections of the installations are normally required for all sizes of building.

District Networks

Although HVAC is executed in individual buildings or other enclosed spaces (like NORAD's underground headquarters), the equipment involved is in some cases an extension of a larger district heating (DH) or district cooling (DC) network, or a combined DHC network. In such cases, the operating and maintenance aspects are simplified and metering becomes necessary to bill for the energy that is consumed, and in some cases energy that is returned to the larger system. For example, at a given time one building may be utilizing chilled water for air conditioning and the warm water it returns may be used in another building for heating, or for the overall heating-portion of the DHC network (likely with energy added to boost the temperature).

Basing HVAC on a larger network helps to provide an economy of scale that is often not possible for individual buildings, for utilizing renewable energy sources such as solar heat, winter's cold, the cooling potential in some places of lakes or seawater for free cooling, and the enabling function of seasonal thermal energy storage.

History

HVAC is based on inventions and discoveries made by Nikolay Lvov, Michael Faraday, Willis Carrier, Edwin Ruud, Reuben Trane, James Joule, William Rankine, Sadi Carnot, and many others.

Multiple inventions within this time frame preceded the beginnings of first comfort air conditioning system, which was designed in 1902 by Alfred Wolff (Cooper, 2003) for the New York Stock Exchange, while Willis Carrier equipped the Sacketts-Wilhems Printing Company with the process AC unit the same year.

The invention of the components of HVAC systems went hand-in-hand with the industrial revolution, and new methods of modernization, higher efficiency, and system control are constantly being introduced by companies and inventors worldwide.

Heating

Heaters are appliances whose purpose is to generate heat (i.e. warmth) for the building. This can be done via central heating. Such a system contains a boiler, furnace, or heat pump to heat water, steam, or air in a central location such as a furnace room in a home, or a mechanical room in a large building. The heat can be transferred by convection, conduction, or radiation.

Generation

Central heating unit

Heaters exist for various types of fuel, including solid fuels, liquids, and gases. Another type of heat source is electricity, normally heating ribbons composed of high resistance wire. This principle is also used for baseboard heaters and portable heaters. Electrical heaters are often used as backup or supplemental heat for heat pump systems.

The heat pump gained popularity in the 1950s in Japan and the United States. Heat pumps can extract heat from various sources, such as environmental air, exhaust air from a building, or from the ground. Initially, heat pump HVAC systems were only used in moderate climates, but with improvements in low temperature operation and reduced loads due to more efficient homes, they are increasing in popularity in cooler climates.

Distribution

Water / Steam

In the case of heated water or steam, piping is used to transport the heat to the rooms. Most modern hot water boiler heating systems have a circulator, which is a pump, to move hot water through the distribution system (as opposed to older gravity-fed systems). The heat can be transferred to the surrounding air using radiators, hot water coils (hydro-air), or other heat exchangers. The radiators may be mounted on walls or installed within the floor to produce floor heat.

The use of water as the heat transfer medium is known as hydronics. The heated water can also supply an auxiliary heat exchanger to supply hot water for bathing and washing.

Air

Warm air systems distribute heated air through duct work systems of supply and return air through metal or fiberglass ducts. Many systems use the same ducts to distribute air cooled by an evaporator coil for air conditioning. The air supply is normally filtered through air cleaners to remove dust and pollen particles.

Dangers

The use of furnaces, space heaters, and boilers as a method of indoor heating could result in incomplete combustion and the emission of carbon monoxide, nitrogen oxides, formaldehyde, volatile organic compounds, and other combustion byproducts. Incomplete combustion occurs when there is insufficient oxygen; the inputs are fuels containing various contaminants and the outputs are harmful byproducts, most dangerously carbon monoxide, which is a tasteless and odorless gas with serious adverse health effects.

Without proper ventilation, carbon monoxide can be lethal at concentrations of 1000 ppm (0.1%). However, at several hundred ppm, carbon monoxide exposure induces headaches, fatigue, nausea, and vomiting. Carbon monoxide binds with hemoglobin in the blood, forming carboxyhemoglobin, reducing the blood's ability to transport oxygen. The primary health concerns associated with carbon monoxide exposure are its cardiovascular and neurobehavioral effects. Carbon monoxide can cause atherosclerosis (the hardening of arteries) and can also trigger heart attacks. Neurologically, carbon monoxide exposure reduces hand to eye coordination, vigilance, and continuous performance. It can also affect time discrimination.

Ventilation

Ventilation is the process of changing or replacing air in any space to control temperature or remove any combination of moisture, odors, smoke, heat, dust, airborne bacteria, or carbon dioxide, and to replenish oxygen. Ventilation includes both the exchange of air with the outside as well as circulation of air within the building. It is one of the most important factors for maintaining acceptable indoor air quality in buildings. Methods for ventilating a building may be divided into *mechanical/forced* and *natural* types.

Mechanical or Forced Ventilation

HVAC ventilation exhaust for a 12-story building

Mechanical, or forced, ventilation is provided by an air handler (AHU) and used to control indoor air quality. Excess humidity, odors, and contaminants can often be controlled via dilution or replacement with outside air. However, in humid climates more energy is required to remove excess moisture from ventilation air.

Kitchens and bathrooms typically have mechanical exhausts to control odors and sometimes humidity. Factors in the design of such systems include the flow rate (which is a function of the fan speed and exhaust vent size) and noise level. Direct drive fans are available for many applications, and can reduce maintenance needs.

Ceiling fans and table/floor fans circulate air within a room for the purpose of reducing the perceived temperature by increasing evaporation of perspiration on the skin of the occupants. Because hot air rises, ceiling fans may be used to keep a room warmer in the winter by circulating the warm stratified air from the ceiling to the floor.

Natural Ventilation

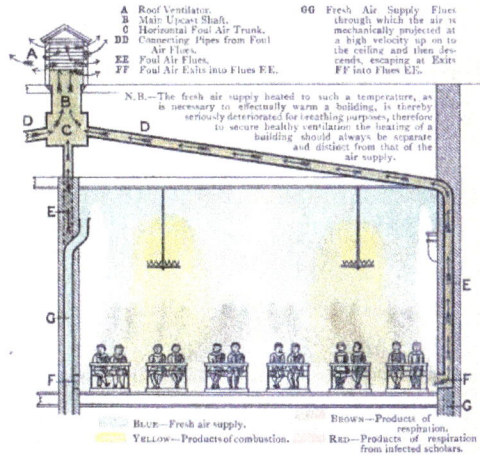

Ventilation on the downdraught system, by impulsion, or the 'plenum' principle, applied to schoolrooms (1899)

Natural ventilation is the ventilation of a building with outside air without using fans or other mechanical systems. It can be via operable windows, louvers, or trickle vents when spaces are small and the architecture permits. In more complex schemes, warm air is allowed to rise and flow out high building openings to the outside (stack effect), causing cool outside air to be drawn into low building openings. Natural ventilation schemes can use very little energy, but care must be taken to ensure comfort. In warm or humid climates, maintaining thermal comfort solely via natural ventilation might not be possible. Air conditioning systems are used, either as backups or supplements. Air-side economizers also use outside air to condition spaces, but do so using fans, ducts, dampers, and control systems to introduce and distribute cool outdoor air when appropriate.

An important component of natural ventilation is air change rate or air changes per hour: the hourly rate of ventilation divided by the volume of the space. For example, six air changes per hour means an amount of new air, equal to the volume of the space, is added every ten minutes. For human comfort, a minimum of four air changes per hour is typical, though warehouses might have only two. Too high of an air change rate may be uncomfortable, akin to a wind tunnel which have thousands of changes per hour. The highest air change rates are for crowded spaces, bars, night clubs, commercial kitchens at around 30 to 50 air changes per hour.

Room pressure can be either positive or negative with respect to outside the room. Positive pressure occurs when there is more air being supplied than exhausted, and is common to reduce the infiltration of outside contaminants.

Airborne Diseases

Natural ventilation is a key factor in reducing the spread of airborne illnesses such as tuberculosis, the common cold, influenza and meningitis. Opening doors, windows, and using ceiling fans are all ways to maximize natural ventilation and reduce the risk of airborne contagion. Natural ventilation requires little maintenance and is inexpensive.

Air Conditioning

An air conditioning system, or a standalone air conditioner, provides cooling and humidity control for all or part of a building. Air conditioned buildings often have sealed windows, because open windows would work against the system intended to maintain constant indoor air conditions. Outside, fresh air is generally drawn into the system by a vent into the indoor heat exchanger section, creating positive air pressure. The percentage of return air made up of fresh air can usually be manipulated by adjusting the opening of this vent. Typical fresh air intake is about 10%.

Air conditioning and refrigeration are provided through the removal of heat. Heat can be removed through radiation, convection, or conduction. Refrigeration conduction media such as water, air, ice, and chemicals are referred to as refrigerants. A refrigerant is employed either in a heat pump system in which a compressor is used to drive thermodynamic refrigeration cycle, or in a free cooling system which uses pumps to circulate a cool refrigerant (typically water or a glycol mix).

Refrigeration Cycle

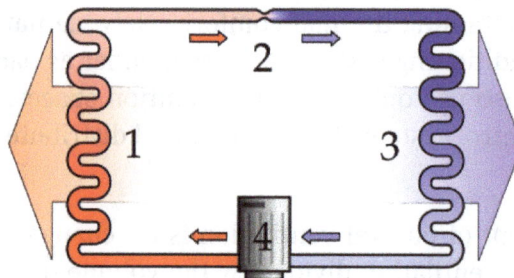

A simple stylized diagram of the refrigeration cycle: 1) condensing coil, 2) expansion valve, 3) evaporator coil, 4) compressor

The refrigeration cycle uses four essential elements to cool.

- The system refrigerant starts its cycle in a gaseous state. The compressor pumps the refrigerant gas up to a high pressure and temperature.

- From there it enters a heat exchanger (sometimes called a condensing coil or condenser) where it loses energy (heat) to the outside, cools, and condenses into its liquid phase.

- An expansion valve (also called metering device) regulates the refrigerant liquid to flow at the proper rate.

- The liquid refrigerant is returned to another heat exchanger where it is allowed to evaporate, hence the heat exchanger is often called an evaporating coil or evaporator. As the liquid refrigerant evaporates it absorbs energy (heat) from the inside air, returns to the compressor, and repeats the cycle. In the process, heat is absorbed from indoors and transferred outdoors, resulting in cooling of the building.

In variable climates, the system may include a reversing valve that switches from heating in winter to cooling in summer. By reversing the flow of refrigerant, the heat pump refrigeration cycle is changed from cooling to heating or vice versa. This allows a facility to be heated and cooled by a single piece of equipment by the same means, and with the same hardware.

Free Cooling

Free cooling systems can have very high efficiencies, and are sometimes combined with seasonal thermal energy storage so the cold of winter can be used for summer air conditioning. Common storage mediums are deep aquifers or a natural underground rock mass accessed via a cluster of small-diameter, heat-exchanger-equipped boreholes. Some systems with small storages are hybrids, using free cooling early in the cooling season, and later employing a heat pump to chill the circulation coming from the storage. The heat pump is added-in because the storage acts as a heat sink when the system is in cooling (as opposed to charging) mode, causing the temperature to gradually increase during the cooling season.

Some systems include an "economizer mode", which is sometimes called a "free-cooling mode". When economizing, the control system will open (fully or partially) the outside air damper and close (fully or partially) the return air damper. This will cause fresh, outside air to be supplied to the system. When the outside air is cooler than the demanded cool air, this will allow the demand to be met without using the mechanical supply of cooling (typically chilled water or a direct expansion "DX" unit), thus saving energy. The control system can compare the temperature of the outside air vs. return air, or it can compare the enthalpy of the air, as is frequently done in climates where humidity is more of an issue. In both cases, the outside air must be less energetic than the return air for the system to enter the economizer mode.

Central vs. Split System

Central, "all-air" air-conditioning systems (or package systems) with a combined outdoor condenser/evaporator unit are often installed in modern residences, offices, and public buildings, but are difficult to retrofit (install in a building that was not designed to receive it) because of the bulky air ducts required. (Minisplit ductless systems are used in these situations.)

An alternative to central systems is the use of separate indoor and outdoor coils in split systems. These systems, although most often seen in residential applications, are gaining popularity in small commercial buildings. The evaporator coil is connected to a remote condenser unit using refrigerant piping between an indoor and outdoor unit instead of ducting air directly from the outdoor unit. Indoor units with directional vents mount onto walls, suspended from ceilings, or fit into the ceiling. Other indoor units mount inside the ceiling cavity, so that short lengths of duct handle air from the indoor unit to vents or diffusers around the rooms.

Dehumidification

Dehumidification (air drying) in an air conditioning system is provided by the evaporator. Since the evaporator operates at a temperature below the dew point, moisture in the air condenses on the evaporator coil tubes. This moisture is collected at the bottom of the evaporator in a pan and removed by piping to a central drain or onto the ground outside.

A dehumidifier is an air-conditioner-like device that controls the humidity of a room or building. It is often employed in basements which have a higher relative humidity because of their lower temperature (and propensity for damp floors and walls). In food retailing establishments, large open chiller cabinets are highly effective at dehumidifying the internal air. Conversely, a humidifier increases the humidity of a building.

Maintenance

All modern air conditioning systems, even small window package units, are equipped with internal air filters. These are generally of a lightweight gauzy material, and must be replaced or washed as conditions warrant. For example, a building in a high dust environment, or a home with furry pets, will need to have the filters changed more often than buildings without these dirt loads. Failure to replace these filters as needed will contribute to a lower heat exchange rate, resulting in wasted energy, shortened equipment life, and higher energy bills; low air flow can result in iced-over evaporator coils, which can completely stop air flow. Additionally, very dirty or plugged filters can cause overheating during a heating cycle, and can result in damage to the system or even fire.

Because an air conditioner moves heat between the indoor coil and the outdoor coil, both must be kept clean. This means that, in addition to replacing the air filter at the evaporator coil, it is also necessary to regularly clean the condenser coil. Failure to keep the condenser clean will eventually result in harm to the compressor, because the condenser coil is responsible for discharging both the indoor heat (as picked up by the evaporator) and the heat generated by the electric motor driving the compressor.

Energy Efficiency

Since the 1980s, manufacturers of HVAC equipment have been making an effort to make the systems they manufacture more efficient. This was originally driven by rising energy costs, and has more recently been driven by increased awareness of environmental issues. Additionally, improvements to the HVAC system efficiency can also help increase occupant health and productivity. In the US, the EPA has imposed tighter restrictions over the years. There are several methods for making HVAC systems more efficient.

Heating Energy

In the past, water heating was more efficient for heating buildings and was the standard in the United States. Today, forced air systems can double for air conditioning and are more popular.

Some benefits of forced air systems, which are now widely used in churches, schools and high-end residences, are

- Better air conditioning effects

- Energy savings of up to 15-20%

- Even conditioning

A drawback is the installation cost, which can be slightly higher than traditional HVAC systems.

Energy efficiency can be improved even more in central heating systems by introducing zoned heating. This allows a more granular application of heat, similar to non-central heating systems. Zones are controlled by multiple thermostats. In water heating systems the thermostats control zone valves, and in forced air systems they control zone dampers inside the vents which selectively block the flow of air. In this case, the control system is very critical to maintaining a proper temperature.

Forecasting is another method of controlling building heating by calculating demand for heating energy that should be supplied to the building in each time unit.

Ground Source Heat Pump

Ground source, or geothermal, heat pumps are similar to ordinary heat pumps, but instead of transferring heat to or from outside air, they rely on the stable, even temperature of the earth to provide heating and air conditioning. Many parts of the country experience seasonal temperature extremes, which would require large-capacity heating and cooling equipment to heat or cool buildings. For example, a conventional heat pump system used to heat a building in Montana's −70 °F (−57 °C) low temperature or cool a building in the highest temperature ever recorded in the US—134 °F (57 °C) in Death Valley, California, in 1913 would require a large amount of energy due to the extreme difference between inside and outside air temperatures. A few feet below the earth's surface, however, the ground remains at a relatively constant temperature. Utilizing this large source of relatively moderate temperature earth, a heating or cooling system's capacity can often be significantly reduced. Although ground temperatures vary according to latitude, at 6 feet (1.8 m) underground, temperatures generally only range from 45 to 75 °F (7 to 24 °C).

An example of a geothermal heat pump that uses a body of water as the heat sink, is the system used by the Trump International Hotel and Tower in Chicago, Illinois. This building is situated on the Chicago River, and uses cold river water by pumping it into a recirculating cooling system, where heat exchangers transfer heat from the building into the water, and then the now-warmed water is pumped back into the Chicago River.

While they may be more costly to install than regular heat pumps, geothermal heat pumps can produce markedly lower energy bills – 30 to 40 percent lower, according to estimates from the US Environmental Protection Agency.

Ventilation Energy Recovery

Energy recovery systems sometimes utilize heat recovery ventilation or energy recovery ventilation systems that employ heat exchangers or enthalpy wheels to recover sensible or latent heat from exhausted air. This is done by transfer of energy to the incoming outside fresh air.

Air Conditioning Energy

The performance of vapor compression refrigeration cycles is limited by thermodynamics. These air conditioning and heat pump devices *move* heat rather than convert it from one form to another, so *thermal efficiencies* do not appropriately describe the performance of these devices. The Coefficient-of-Performance (COP) measures performance, but this dimensionless measure has not been adopted. Instead, the Energy Efficiency Ratio (*EER*) has traditionally been used to characterize the performance of many HVAC systems. EER is the Energy Efficiency Ratio based on a 35 °C (95 °F) outdoor temperature. To more accurately describe the performance of air conditioning equipment over a typical cooling season a modified version of the EER, the Seasonal Energy Efficiency Ratio (*SEER*), or in Europe the ESEER, is used. SEER ratings are based on seasonal temperature averages instead of a constant 35 °C (95 °F) outdoor temperature. The current industry minimum SEER rating is 14 SEER.

Engineers have pointed out some areas where efficiency of the existing hardware could be improved. For example, the fan blades used to move the air are usually stamped from sheet metal, an economical method of manufacture, but as a result they are not aerodynamically efficient. A well-designed blade could reduce electrical power required to move the air by a third.

Air Filtration and Cleaning

Air handling unit, used for heating, cooling, and filtering the air

Air cleaning and filtration removes particles, contaminants, vapors and gases from the air. The filtered and cleaned air then is used in heating, ventilation and air conditioning. Air cleaning and filtration should be taken in account when protecting our building environments.

Clean Air Delivery Rate and Filter Performance

Clean air delivery rate is the amount of clean air an air cleaner provides to a room or space. When determining CADR, the amount of airflow in a space is taken into account. For example, an air cleaner with a flow rate of 100 cfm (cubic feet per minute) and an efficiency of 50% has a CADR of 50 cfm. Along with CADR, filtration performance is very important when it comes to the air in our indoor environment. Filter performance depends on the size of the particle or fiber, the filter packing density and depth and also the air flow rate.

HVAC Industry and Standards

The HVAC industry is a worldwide enterprise, with roles including operation and maintenance, system design and construction, equipment manufacturing and sales, and in education and research. The HVAC industry was historically regulated by the manufacturers of HVAC equipment, but regulating and standards organizations such as HARDI, ASHRAE, SMACNA, ACCA, Uniform Mechanical Code, International Mechanical Code, and AMCA have been established to support the industry and encourage high standards and achievement.

The starting point in carrying out an estimate both for cooling and heating depends on the exterior climate and interior specified conditions. However, before taking up the heat load calculation, it is necessary to find fresh air requirements for each area in detail, as pressurization is an important consideration.

International

ISO 16813:2006 is one of the ISO building environment standards. It establishes the general principles of building environment design. It takes into account the need to provide a healthy indoor environment for the occupants as well as the need to protect the environment for future generations and promote collaboration among the various parties involved in building environmental design for sustainability. ISO16813 is applicable to new construction and the retrofit of existing buildings.

The building environmental design standard aims to:

- provide the constraints concerning sustainability issues from the initial stage of the design process, with building and plant life cycle to be considered together with owning and operating costs from the beginning of the design process;

- assess the proposed design with rational criteria for indoor air quality, thermal comfort, acoustical comfort, visual comfort, energy efficiency and HVAC system controls at every stage of the design process;

- iterate decisions and evaluations of the design throughout the design process.

North America

United States

In the United States, HVAC engineers generally are members of the American Society of Heating, Refrigerating, and Air-Conditioning Engineers (ASHRAE), EPA Universal CFC certified, or locally engineer certified such as a Special to Chief Boilers License issued by the state or, in some jurisdictions, the city. ASHRAE is an international technical society for all individuals and organizations interested in HVAC. The Society, organized into regions, chapters, and student branches, allows exchange of HVAC knowledge and experiences for the benefit of the field's practitioners and the public. ASHRAE provides many opportunities to participate in the development of new knowledge via, for example, research and its many technical committees. These committees typically meet twice per year at the ASHRAE Annual and Winter Meetings. A popular product show, the AHR Expo, is held in conjunction with each winter meeting. The Society has approximately 50,000 members and has headquarters in Atlanta, Georgia.

The most recognized standards for HVAC design are based on ASHRAE data. The most general of four volumes of the ASHRAE Handbook is Fundamentals; it includes heating and cooling calculations. Each volume of the ASHRAE Handbook is updated every four years. The design professional must consult ASHRAE data for the standards of design and care as the typical building codes provide little to no information on HVAC design practices; codes such as the UMC and IMC do include much detail on installation requirements, however. Other useful reference materials include items from SMACNA, ACGIH, and technical trade journals.

American design standards are legislated in the Uniform Mechanical Code or International Mechanical Code. In certain states, counties, or cities, either of these codes may be adopted and amended via various legislative processes. These codes are updated and published by the International Association of Plumbing and Mechanical Officials (IAPMO) or the International Code Council (ICC) respectively, on a 3-year code development cycle. Typically, local building permit departments are charged with enforcement of these standards on private and certain public properties.

HVAC professionals in the US can receive training through formal training institutions, where most earn associate degrees. Training for HVAC technicians includes classroom lectures and hands-on tasks, and can be followed by an apprenticeship wherein the recent graduate works alongside a professional HVAC technician for a temporary period. HVAC techs who have been trained can also be certified in areas such as air conditioning, heat pumps, gas heating, and commercial refrigeration.

Europe

United Kingdom

The Chartered Institution of Building Services Engineers is a body that covers the essential Service (systems architecture) that allow buildings to operate. It includes the electrotechnical, heating, ventilating, air conditioning, refrigeration and plumbing industries. To train as a building services engineer, the academic requirements are GCSEs (A-C) / Standard Grades (1-3) in Maths and Science, which are important in measurements, planning and theory. Employers will often want a degree in a branch of engineering, such as building environment engineering, electrical engineering or mechanical engineering. To become a full member of CIBSE, and so also to be registered by the Engineering Council UK as a chartered engineer, engineers must also attain an Honours Degree and a master's degree in a relevant engineering subject.

CIBSE publishes several guides to HVAC design relevant to the UK market, and also the Republic of Ireland, Australia, New Zealand and Hong Kong. These guides include various recommended design criteria and standards, some of which are cited within the UK building regulations, and therefore form a legislative requirement for major building services works. The main guides are:

- Guide A: Environmental Design
- Guide B: Heating, Ventilating, Air Conditioning and Refrigeration
- Guide C: Reference Data
- Guide D: Transportation systems in Buildings

- Guide E: Fire Safety Engineering

- Guide F: Energy Efficiency in Buildings

- Guide G: Public Health Engineering

- Guide H: Building Control Systems

- Guide J: Weather, Solar and Illuminance Data

- Guide K: Electricity in Buildings

- Guide L: Sustainability

- Guide M: Maintenance Engineering and Management

Within the construction sector, it is the job of the building services engineer to design and oversee the installation and maintenance of the essential services such as gas, electricity, water, heating and lighting, as well as many others. These all help to make buildings comfortable and healthy places to live and work in. Building Services is part of a sector that has over 51,000 businesses and employs represents 2%-3% of the GDP.

Australia

The Air Conditioning and Mechanical Contractors Association of Australia (AMCA), Australian Institute of Refrigeration, Air Conditioning and Heating (AIRAH), and CIBSE are responsible.

Asia

Asian architectural temperature-control have different priorities than European methods. For example, Asian heating traditionally focuses on maintaining temperatures of objects such as the floor or furnishings such as Kotatsu tables and directly warming people, as opposed to the Western focus, in modern periods, on designing air systems.

Philippines

The Philippine Society of Ventilating, Air Conditioning and Refrigerating Engineers (PSVARE) along with Philippine Society of Mechanical Engineers (PSME) govern on the codes and standards for HVAC / MVAC in the Philippines.

India

The Indian Society of Heating, Refrigerating and Air Conditioning Engineers (ISHRAE) was established to promote the HVAC industry in India. ISHRAE is an associate of ASHRAE. ISHRAE was started at Delhi in 1981 and a chapter was started in Bangalore in 1989. Between 1989 & 1993, ISHRAE chapters were formed in all major cities in India and also in the Middle East.

Air Conditioning

Air conditioning units outside a building

Window unit inside a room

Air conditioning (often referred to as AC, A.C., or A/C) is the process of removing heat from a confined space, thus cooling the air, and removing humidity. This process is used to achieve a more comfortable interior environment, typically for humans or animals; however, air conditioning is also used to cool/dehumidify rooms filled with heat-producing electronic devices, such as computer servers power amplifiers, and even to display and store artwork. Air conditioners often use a fan to distribute the conditioned air to an occupied space such as a building or a car to improve thermal comfort and indoor air quality. Electric refrigerant-based AC units range from small units that can cool a small bedroom, which can be carried by a single adult to massive units installed on the roof of office towers that can cool an entire building. The cooling is typically achieved through a refrigeration cycle, but sometimes evaporation or free cooling is used. Air conditioning systems can also be made based on desiccants. (chemicals which remove moisture from the air) and subterraneous pipes that can distribute the heated refrigerant to the ground for cooling.

In the most general sense, air conditioning can refer to any form of technology that modifies the condition of air (heating, cooling, (de-)humidification, cleaning, ventilation, or air movement). In common usage, though, "air conditioning" refers to systems which cool air. In construction, a complete system of heating, ventilation, and air conditioning is referred to as heating, ventilation, and air conditioning (HVAC – as opposed to *AC*).

History

Evaporative Cooling

Since prehistoric times, snow and ice were used for cooling. The business of harvesting ice during winter and storing for use in summer became popular towards the late 17th century. This practice was replaced by mechanical ice-making machines.

The basic concept behind air conditioning is said to have been applied in ancient Egypt, where reeds were hung in windows and were moistened with trickling water. The evaporation of water cooled the air blowing through the window. This process also made the air more humid, which can be beneficial in a dry desert climate. In Ancient Rome, water from aqueducts was circulated through the walls of certain houses to cool them. Other techniques in medieval Persia involved the use of cisterns and wind towers to cool buildings during the hot season.

The 2nd-century Chinese inventor Ding Huan (fl 180) of the Han Dynasty invented a rotary fan for air conditioning, with seven wheels 3 m (10 ft) in diameter and manually powered by prisoners of the time. In 747, Emperor Xuanzong (r. 712–762) of the Tang Dynasty (618–907) had the *Cool Hall* (*Liang Tian*) built in the imperial palace, which the *Tang Yulin* describes as having water-powered fan wheels for air conditioning as well as rising jet streams of water from fountains. During the subsequent Song Dynasty (960–1279), written sources mentioned the air conditioning rotary fan as even more widely used.

In the 17th century, Cornelis Drebbel demonstrated "Turning Summer into Winter" for James I of England by adding salt to water.

Development of Mechanical Cooling

Modern air conditioning emerged from advances in chemistry during the 19th century, and the first large-scale electrical air conditioning was invented and used in 1902 by American inventor Willis Carrier. The introduction of residential air conditioning in the 1920s helped enable the great migration to the Sun Belt in the United States.

In 1758, Benjamin Franklin and John Hadley, a chemistry professor at Cambridge University, conducted an experiment to explore the principle of evaporation as a means to rapidly cool an object. Franklin and Hadley confirmed that evaporation of highly volatile liquids (such as alcohol and ether) could be used to drive down the temperature of an object past the freezing point of water. They conducted their experiment with the bulb of a mercury thermometer as their object and with a bellows used to speed-up the evaporation. They lowered the temperature of the thermometer bulb down to −14 °C (7 °F) while the ambient temperature was 18 °C (64 °F). Franklin noted that, soon after they passed the freezing point of water 0 °C (32 °F), a thin film of ice formed on the surface of the thermometer's bulb and that the ice mass was about 6 mm ($\frac{1}{4}$ in) thick when they stopped the experiment upon reaching −14 °C (7 °F). Franklin concluded: «From this experiment one may see the possibility of freezing a man to death on a warm summer›s day»

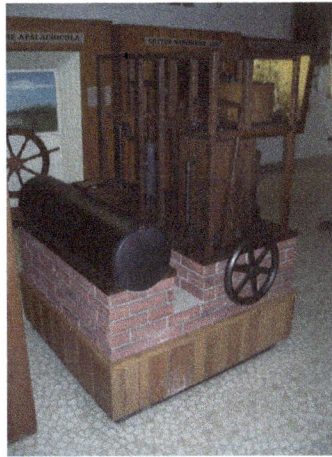

Three-quarters scale model of Gorrie's ice machine John Gorrie State Museum, Florida

In 1820, English scientist and inventor Michael Faraday discovered that compressing and liquefying ammonia could chill air when the liquefied ammonia was allowed to evaporate. In 1842, Florida physician John Gorrie used compressor technology to create ice, which he used to cool air for his patients in his hospital in Apalachicola, Florida. He hoped to eventually use his ice-making machine to regulate the temperature of buildings. He even envisioned centralized air conditioning that could cool entire cities. Though his prototype leaked and performed irregularly, Gorrie was granted a patent in 1851 for his ice-making machine. Improved process for the artificial production of ice. His hopes for its success vanished soon afterwards when his chief financial backer died; Gorrie did not get the money he needed to develop the machine. According to his biographer, Vivian M. Sherlock, he blamed the "Ice King", Frederic Tudor, for his failure, suspecting that Tudor had launched a smear campaign against his invention. Dr. Gorrie died impoverished in 1855, and the idea of air conditioning went away for 50 years.

James Harrison's first mechanical ice-making machine began operation in 1851 on the banks of the Barwon River at Rocky Point in Geelong (Australia). His first commercial ice-making machine followed in 1853, and his patent for an ether vapor compression refrigeration system was granted in 1855. This novel system used a compressor to force the refrigeration gas to pass through a condenser, where it cooled down and liquefied.

The liquefied gas then circulated through the refrigeration coils and vaporized again, cooling down the surrounding system. The machine employed a flywheel and produced 3,000 kilograms of ice per day.

Though Harrison had commercial success establishing a second ice company back in Sydney in 1860, he later entered the debate over how to compete against the American advantage of unrefrigerated beef sales to the United Kingdom. He wrote: "Fresh meat frozen and packed as if for a voyage, so that the refrigerating process may be continued for any required period", and in 1873 prepared the sailing ship Norfolk for an experimental beef shipment to the United Kingdom. His choice of a cold room system instead of installing a refrigeration system upon the ship itself proved disastrous when the ice was consumed faster than expected.

Electromechanical Cooling

Willis Carrier

In 1902, the first modern electrical air conditioning unit was invented by Willis Carrier in Buffalo, New York. After graduating from Cornell University, Carrier found a job at the Buffalo Forge Company. While there, he began experimenting with air conditioning as a way to solve an application problem for the Sackett-Wilhelms Lithographing and Publishing Company in Brooklyn, New York. The first air conditioner, designed and built in Buffalo by Carrier, began working on 17 July 1902.

Designed to improve manufacturing process control in a printing plant, Carrier's invention controlled not only temperature but also humidity. Carrier used his knowledge of the heating of objects with steam and reversed the process. Instead of sending air through hot coils, he sent it through cold coils (filled with cold water). The air was cooled, and thereby the amount of moisture in the air could be controlled, which in turn made the humidity in the room controllable. The controlled temperature and humidity helped maintain consistent paper dimensions and ink alignment. Later, Carrier's technology was applied to increase productivity in the workplace, and The Carrier Air Conditioning Company of America was formed to meet rising demand. Over time, air conditioning came to be used to improve comfort in homes and automobiles as well. Residential sales expanded dramatically in the 1950s.

In 1906, Stuart W. Cramer of Charlotte, North Carolina was exploring ways to add moisture to the air in his textile mill. Cramer coined the term "air conditioning", using it in a patent claim he filed that year as an analogue to "water conditioning", then a well-known process for making textiles easier to process. He combined moisture with ventilation to "condition" and change the air in the factories, controlling the humidity so necessary in textile plants. Willis Carrier adopted the term and incorporated it into the name of his company.

Shortly thereafter, the first private home to have air conditioning was built in Minneapolis in 1914, owned by Charles Gates. Realizing that air conditioning would one day be a standard feature of private homes, particularly in regions with warmer climate, David St. Pierre DuBose (1898-1994) designed a network of ductwork and vents for his home *Meadowmont*, all disguised behind intricate and attractive Georgian-style open moldings.This building is believed to be one of the first private homes in the United States equipped for central air conditioning.

In 1945, Robert Sherman of Lynn, Massachusetts invented a portable, in-window air conditioner that cooled, heated, humidified, dehumidified, and filtered the air.

Refrigerant Development

A modern R-134a hermetic refrigeration compressor

The first air conditioners and refrigerators employed toxic or flammable gases, such as ammonia, methyl chloride, or propane, that could result in fatal accidents when they leaked. Thomas Midgley, Jr. created the first non-flammable, non-toxic chlorofluorocarbon gas, *Freon*, in 1928. The name is a trademark name owned by DuPont for any chlorofluorocarbon (CFC), hydrochlorofluorocarbon (HCFC), or hydrofluorocarbon (HFC) refrigerant. The refrigerant names include a number indicating the molecular composition (e.g., R-11, R-12, R-22, R-134A). The blend most used in direct-expansion home and building comfort cooling is an HCFC known as chlorodifluoromethane (R-22).

Dichlorodifluoromethane (R-12) was the most common blend used in automobiles in the US until 1994, when most designs changed to R-134A due to the ozone-depleting potential of R-12. R-11 and R-12 are no longer manufactured in the US for this type of application, so the only source for air-conditioning repair purposes is the cleaned and purified gas recovered from other air conditioner systems. Several non-ozone-depleting refrigerants have been developed as alternatives, including R-410A. It was first commercially used by Carrier Corp. under the brand name *Puron*.

Modern refrigerants have been developed to be more environmentally safe than many of the early chlorofluorocarbon-based refrigerants used in the early- and mid-twentieth century. These include HCFCs (R-22, as used in most U.S. homes even before 2011) and HFCs (R-134a, used in most cars) have replaced most CFC use. HCFCs, in turn, are supposed to have been in the process of being phased out under the Montreal Protocol and replaced by HFCs such as R-410A, which lack chlorine. HFCs, however, contribute to climate change problems. Moreover, policy and political influence by corporate executives resisted change. Corporations insisted that no alternatives to HFCs existed. The environmental organization Greenpeace solicited a European laboratory to research an alternative ozone- and climate-safe refrigerant in 1992, gained patent rights to a hydrocarbon mix of isopentane and isobutane, but then left the technology as open access. Their activist marketing first in Germany led to companies like Whirlpool, Bosch, and later LG and others to incorporate the technology throughout Europe, then Asia, although the corporate executives resisted in Latin America, so that it arrived in Argentina produced by a domestic firm in 2003, and then finally with giant Bosch's production in Brazil by 2004. In 1995, Germany made CFC refrigerators illegal. DuPont and other companies blocked the refrigerant in the U.S. with the U.S. E.P.A., disparaging the approach as "that German technology." Nevertheless, in 2004, Greenpeace worked with multinational corporations like Coca-Cola and Unilever, and later Pepsico and others, to create a corporate coalition called Refrigerants Naturally!. Then, four years later, Ben & Jerry's of Unilever and General Electric began to take steps to support production and use in the U.S. Only in 2011 did the E.P.A. finally decide in favor of the ozone- and climate-safe refrigerant for U.S. manufacture.

Operating Principles

Refrigeration Cycle

In the refrigeration cycle, heat is transported from a colder location to a hotter area. As heat would naturally flow in the opposite direction, work is required to achieve this. A refrigerator is an example of such a system, as it transports the heat out of the interior and into its environment. The refrigerant is used as the medium which absorbs and removes heat from the space to be cooled and subsequently rejects that heat elsewhere.

Capillary expansion valve connection to evaporator inlet. Notice frost formation

Circulating refrigerant vapor enters the compressor, where its pressure and temperature are increased. The hot, compressed refrigerant vapor is now at a temperature and pressure at which it can be condensed and is routed through a condenser. Here it is cooled by air flowing across the condenser coils and condensed into a liquid. Thus, the circulating refrigerant removes heat from the system and the heat is carried away by the air. The removal of this heat can be greatly augmented by pouring water over the condenser coils, making it much cooler when it hits the expansion valve.

The condensed, pressurized, and still usually somewhat hot liquid refrigerant is next routed through an expansion valve (often nothing more than a pinhole in the system's copper tubing) where it undergoes an abrupt reduction in pressure. That pressure reduction results in flash evaporation of a part of the liquid refrigerant, greatly lowering its temperature. The cold refrigerant is then routed through the evaporator. A fan blows the interior warm air (which is to be cooled) across the evaporator, causing the liquid part of the cold refrigerant mixture to evaporate as well, further lowering the temperature. The warm air is therefore cooled and is pumped by an exhaust fan/ blower into the room.

To complete the refrigeration cycle, the refrigerant vapor is routed back into the compressor. In order for the process to have any efficiency, the cooling/ evaporative portion of the system must be separated by some kind of physical barrier from the heating/ condensing portion, and each portion must have its own fan to circulate its own "kind" of air (either the hot air or the cool air). Modern air conditioning systems are not designed to draw air into the room from the outside, they only recirculate the increasingly cool air on the inside. Because this inside air always has some amount of moisture suspended in it, the cooling portion of the process always causes ambient warm water vapor to condense on the cooling coils and to drip from them down onto a catch tray at the bottom of the unit from which it must then be routed outside, usually through a drain hole.

As this moisture has no dissolved minerals in it, it never causes mineral buildup on the coils, though if the unit is set at its strongest cooling setting and happens to have inadequate circulation of air through the coils and also experiences a failure of the thermistor which senses the ambient temperature in the room, the coil's fins can develop a layer of ice which will then grow and eventually block the circulation of air on the cool side of the unit altogether in a positive feedback loop that will cause the formation of an ice block inside the unit: only minuscule amounts of cool air will then manage to come from the exhaust vent until this ice is removed or is allowed to melt. This will happen even if the ambient humidity level is low: once ice begins to form on the evaporative fins, it will reduce circulation efficiency and cause the development of more ice, etc. A clean and strong circulatory fan can help prevent this, as will raising the target cool temperature of the unit's thermostat to a point that the compressor is allowed to turn off occasionally. A failing thermistor may also cause this problem. This is the same issue faced by refrigerators that do not have a defrost cycle. Dust can also cause the fins to begin blocking air flow with the same undesirable result: ice.

By running an air conditioner's compressor in the opposite direction, the overall effect can be completely reversed and the indoor compartment will become heated instead of cooled. The engineering of physical and thermodynamic properties of gas–vapor mixtures is called psychrometrics.

Heat Pump Unit

An example of an externally fitted AC unit which uses a heat pump system.

A heat pump is an air conditioner in which the refrigeration cycle can be reversed, producing heating instead of cooling in the indoor environment. They are also commonly referred to as a "reverse cycle air conditioner". The heat pump is significantly more energy efficient than electric resistance heating. Some homeowners elect to have a heat pump system installed as a feature of a central air conditioner. When the heat pump is in heating mode, the indoor evaporator coil switches roles and becomes the condenser coil, producing heat. The outdoor condenser unit also switches roles to serve as the evaporator, and discharges cold air (colder than the ambient outdoor air).

Air-source heat pumps are more popular in milder winter climates where the temperature is frequently in the range of 40–55 °F (4–13 °C), because heat pumps become inefficient in more extreme cold. This is because ice forms on the outdoor unit's heat exchanger coil, which blocks air flow over the coil. To compensate for this, the heat pump system must temporarily switch back into the regular air conditioning mode to switch the outdoor evaporator coil *back* to being the condenser coil, so that it can heat up and defrost. A heat pump system will therefore have a form of electric resistance heating in the indoor air path that is activated only in this mode in order to compensate for the temporary indoor air cooling, which would otherwise be uncomfortable in the winter. The icing problem becomes much more severe with lower outdoor temperatures, so heat pumps are commonly installed in tandem with a more conventional form of heating, such as a natural gas or oil furnace, which is used instead of the heat pump during harsher winter temperatures. In this case, the heat pump is used efficiently during the milder temperatures, and the system is switched to the conventional heat source when the outdoor temperature is lower.

Absorption heat pumps are a kind of air-source heat pump, but they do not depend on electricity to power them. Instead, gas, solar power, or heated water is used as a main power source. An absorption pump dissolves ammonia gas in water, which gives off heat. Next, the water and ammonia mixture is depressurized to induce boiling, and the ammonia is boiled off, which absorbs heat from the outdoor air.

Some more expensive fixed window air conditioning units have a true heat pump function. However, a window unit may only have an electric resistance heater.

Evaporative Cooling

An evaporative cooler

In very dry climates, evaporative coolers, sometimes referred to as swamp coolers or desert coolers, are popular for improving coolness during hot weather. An evaporative cooler is a device that draws outside air through a wet pad, such as a large sponge soaked with water. The sensible heat of the incoming air, as measured by a dry bulb thermometer, is reduced. The temperature of the incoming air is reduced, but it is also more humid, so the total heat (sensible heat plus latent heat) is unchanged. Some of the sensible heat of the entering air is converted to latent heat by the evaporation of water in the wet cooler pads. If the entering air is dry enough, the results can be quite substantial.

Evaporative coolers tend to feel as if they are not working during times of high humidity, when there is not much dry air with which the coolers can work to make the air as cool as possible for dwelling occupants. Unlike other types of air conditioners, evaporative coolers rely on the outside air to be channeled through cooler pads that cool the air before it reaches the inside of a house through its air duct system; this cooled outside air must be allowed to push the warmer air within the house out through an exhaust opening such as an open door or window. These coolers cost less and are mechanically simple to understand and maintain.

Free Cooling

Air conditioning can also be provided by a process called free cooling which uses pumps to circulate a coolant (typically water or a glycol mix) from a cold source, which in turn acts as a heat sink for the energy that is removed from the cooled space. Common storage media are deep aquifers or a natural underground rock mass accessed via a cluster of small-diameter boreholes, equipped with heat exchanger. Some systems with small storage capacity are hybrid systems, using free cooling early in the cooling season, and later employing a heat pump to chill the circulation coming from the storage. The heat pump is added because the temperature of the storage gradually increases during the cooling season, thereby declining its effectiveness.

Free cooling systems can have very high efficiencies, and are sometimes combined with seasonal thermal energy storage (STES) so the cold of winter can be used for summer air conditioning. Free cooling and hybrid systems are mature technology.

Humidity Control

Since humans perspire to provide natural cooling by the evaporation of perspiration from the skin, drier air (up to a point) improves the comfort provided. The comfort air conditioner is designed to create a 50% to 60% relative humidity in the occupied space.

Dehumidification and Cooling

Refrigeration air conditioning equipment usually reduces the absolute humidity of the air processed by the system. The relatively cold (below the dewpoint) evaporator coil condenses water vapor from the processed air, much like an ice-cold drink will condense water on the outside of a glass. Therefore, water vapor is removed from the cooled air and the relative humidity in the room is lowered. The water is usually sent to a drain or may simply drip onto the ground outdoors. The heat is rejected by the condenser which is located outside of room to be cooled.

Dehumidification Only

Typical portable dehumidifier

A specialized air conditioner that is used only for dehumidifying is called a dehumidifier. It also uses a refrigeration cycle, but differs from a standard air conditioner in that both the evaporator and the condenser are placed in the same air path. A standard air conditioner transfers heat energy out of the room because its condenser coil releases heat outside. However, since all components of the dehumidifier are in the *same* room, no heat energy is removed. Instead, the electric power consumed by the dehumidifier remains in the room as heat, so the room is actually *heated*, just as by an electric heater that draws the same amount of power.

In addition, if water is condensed in the room, the amount of heat previously needed to evaporate that water also is re-released in the room (the latent heat of vaporization). The dehumidification process is the inverse of adding water to the room with an evaporative cooler, and instead releases heat. Therefore, an in-room dehumidifier always will warm the room and reduce the relative humidity indirectly, as well as reducing the humidity directly by condensing and removing water.

Inside the unit, the air passes over the evaporator coil first, and is cooled and dehumidified. The now dehumidified, cold air then passes over the condenser coil where it is warmed up again. Then the air is released back into the room. The unit produces warm, dehumidified air and can usually be placed freely in the environment (room) that is to be conditioned.

Dehumidifiers are commonly used in cold, damp climates to prevent mold growth indoors, especially in basements. They are also used to protect sensitive equipment from the adverse effects of excessive humidity in tropical countries.

Energy Transfer

In a thermodynamically closed system, any power dissipated into the system that is being maintained at a set temperature (which is a standard mode of operation for modern air conditioners) requires that the rate of energy removal by the air conditioner increase. This increase has the effect that, for each unit of energy input into the system (say to power a light bulb in the closed system), the air conditioner removes that energy. In order to do so, the air conditioner must increase its power consumption by the inverse of its "efficiency" (coefficient of performance) times the amount of power dissipated into the system. As an example, assume that inside the closed system a 100 W heating element is activated, and the air conditioner has an coefficient of performance of 200%. The air conditioner's power consumption will increase by 50 W to compensate for this, thus making the 100 W heating element cost a total of 150 W of power.

It is typical for air conditioners to operate at "efficiencies" of significantly greater than 100%. However, it may be noted that the input electrical energy is of higher thermodynamic quality (lower entropy) than the output thermal energy (heat energy).

Air conditioner equipment power in the U.S. is often described in terms of "tons of refrigeration". A ton of refrigeration is approximately equal to the cooling power of one short ton (2000 pounds or 907 kilograms) of ice melting in a 24-hour period. The value is defined as 12,000 BTU per hour, or 3517 watts. Residential central air systems are usually from 1 to 5 tons (3 to 20 kilowatts (kW)) in capacity.

Seasonal Energy Efficiency Ratio

For residential homes, some countries set minimum requirements for energy efficiency. In the United States, the efficiency of air conditioners is often (but not always) rated by the *seasonal energy efficiency ratio (SEER)*. The higher the SEER rating, the more energy efficient is the air conditioner. The SEER rating is the BTU of cooling output during its normal annual usage divided by the total electric energy input in watt hours (W·h) during the same period.

$$SEER = BTU \div (W \cdot h)$$

this can also be rewritten as:

$$SEER = (BTU / h) \div W,$$ where "W" is the average electrical power in Watts, and (BTU/h) is the rated cooling power.

For example, a 5000 BTU/h air-conditioning unit, with a SEER of 10, would consume 5000/10 = 500 Watts of power on average.

The electrical energy consumed per year can be calculated as the average power multiplied by the annual operating time:

$$500 \text{ W} \times 1000 \text{ h} = 500{,}000 \text{ W·h} = 500 \text{ kWh}$$

Assuming 1000 hours of operation during a typical cooling season (i.e., 8 hours per day for 125 days per year).

Another method that yields the same result, is to calculate the total annual cooling output:

$$5000 \text{ BTU/h} \times 1000 \text{ h} = 5{,}000{,}000 \text{ BTU}$$

Then, for a SEER of 10, the annual electrical energy usage would be:

$$5{,}000{,}000 \text{ BTU} \div 10 = 500{,}000 \text{ W·h} = 500 \text{ kWh}$$

SEER is related to the coefficient of performance (COP) commonly used in thermodynamics and also to the Energy Efficiency Ratio (EER). The EER is the efficiency rating for the equipment at a particular pair of external and internal temperatures, while SEER is calculated over a whole range of external temperatures (i.e., the temperature distribution for the geographical location of the SEER test). SEER is unusual in that it is composed of an Imperial unit divided by an SI unit. The COP is a ratio with the same metric units of energy (joules) in both the numerator and denominator. They cancel out, leaving a dimensionless quantity. Formulas for the approximate conversion between SEER and EER or COP are available from the Pacific Gas and Electric Company:

(1) SEER = EER ÷ 0.9

(2) SEER = COP × 3.792

(3) EER = COP × 3.413

From equation (2) above, a SEER of 13 is equivalent to a COP of 3.43, which means that 3.43 units of heat energy are pumped per unit of work energy.

The United States now requires that residential systems manufactured in 2006 have a minimum SEER rating of 13 (although window-box systems are exempt from this law, so their SEER is still around 10).

Installation Types

Window Unit and Packaged Terminal

Window unit air conditioners are installed in an open window. The interior air is cooled as a fan blows it over the evaporator. On the exterior the heat drawn from the interior is dissipated into the environment as a second fan blows outside air over the condenser. A large house or building may have several such units, allowing each room to be cooled separately.

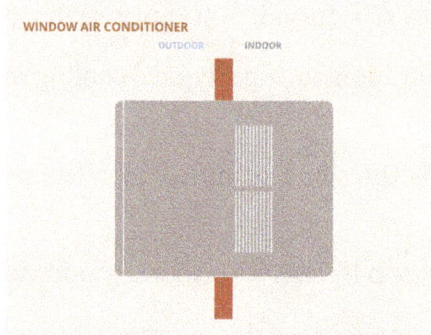

How a window air conditioner works

Air conditioning window unit

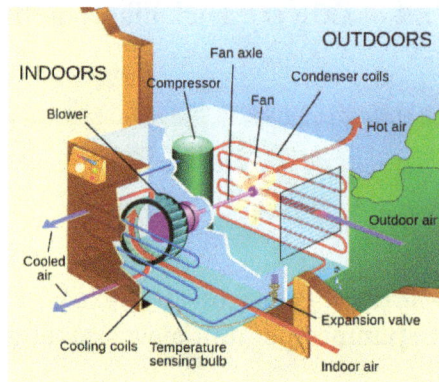

Parts of a window unit

Packaged terminal air conditioner (PTAC) systems are also known as wall-split air conditioning systems. They are ductless systems. PTACs, which are frequently used in hotels, have two separate units (terminal packages), the evaporative unit on the interior and the condensing unit on the exterior, with an opening passing through the wall and connecting them. This minimizes the interior system footprint and allows each room to be adjusted independently. PTAC systems may be adapted to provide heating in cold weather, either directly by using an electric strip, gas, or other heater, or by reversing the refrigerant flow to heat the interior and draw heat from the exterior air, converting the air conditioner into a heat pump. While room air conditioning provides maximum flexibility, when used to cool many rooms at a time it is generally more expensive than central air conditioning.

The first practical through-the-wall air conditioning unit was invented by engineers at Chrysler Motors and offered for sale starting in 1935.

Split Systems

Split-system air conditioners come in two forms: mini-split and central systems. In both types, the inside-environment (evaporative) heat exchanger is separated by some distance from the outside-environment (condensing unit) heat exchanger.

Mini-split (Ductless) System

Outside part of a ductless split-type air conditioner

Indoor part of a ductless split-type air conditioner

A mini-split system typically supplies air conditioned and heated air to a single or a few rooms of a building. Mutli-zone systems are a common application of ductless systems and allow up to 8 rooms (zones) to be conditioned from a single outdoor unit. Multi-zone systems typically offer a variety of indoor unit styles including wall-mounted, ceiling-mounted, ceiling recessed, and horizontal ducted. Mini-split systems typically produce 9,000 to 36,000 Btu (9,500–38,000 kJ) per hour of cooling. Multi-zone systems provide extended cooling and heating capacity up to 60,000 Btu's.

Advantages of the ductless system include smaller size and flexibility for zoning or heating and cooling individual rooms. The inside wall space required is significantly reduced. Also, the compressor and heat exchanger can be located farther away from the inside space, rather than merely on the other side of the same unit as in a PTAC or window air conditioner. Flexible exterior hoses lead from the outside unit to the interior one(s); these are often enclosed with metal to look like common drainpipes from the roof. In addition, ductless systems offer higher efficiency, reaching above 30 SEER.

The primary disadvantage of ductless air conditioners is their cost. Such systems cost about US$1,500 to US$2,000 per ton (12,000 BTU per hour) of cooling capacity. This is about 30% more than central systems (not including ductwork) and may cost more than twice as much as window units of similar capacity."

An additional possible disadvantage that the cost of installing mini splits can be higher than some systems, although lower operating costs and rebates or other financial incentives—offered in some areas—can help offset the initial expense.

Central (Ducted) Air Conditioning

Central (ducted) air conditioning offers whole-house or large-commercial-space cooling, and often offers moderate multi-zone temperature control capability by the addition of air-louver-control boxes.

In central air conditioning, the inside heat-exchanger is typically placed inside the central furnace/AC unit of the forced air heating system which is then used in the summer to distribute chilled air throughout a residence or commercial building.

Portable Units

A portable air conditioner can be easily transported inside a home or office. They are currently available with capacities of about 5,000–60,000 BTU/h (1,500–18,000 W) and with or without electric-resistance heaters. Portable air conditioners are either evaporative or refrigerative.

The compressor-based refrigerant systems are air-cooled, meaning they use air to exchange heat, in the same way as a car radiator or typical household air conditioner does. Such a system dehumidifies the air as it cools it. It collects water condensed from the cooled air and produces hot air which must be vented outside the cooled area; doing so transfers heat from the air in the cooled area to the outside air.

Portable Split System

A portable system has an indoor unit on wheels connected to an outdoor unit via flexible pipes, similar to a permanently fixed installed unit.

Portable Hose System

Hose systems, which can be *monoblock* or *air-to-air*, are vented to the outside via air ducts. The *monoblock* type collects the water in a bucket or tray and stops when full.

The *air-to-air* type re-evaporates the water and discharges it through the ducted hose and can run continuously.

A single-hose unit uses air from within the room to cool its condenser, and then vents it outside. This air is replaced by hot air from outside or other rooms (due to the negative pressure inside the room), thus reducing the unit's overall efficiency.

Modern units might have a coefficient of performance of approximately 3 (i.e., 1 kW of electricity will produce 3 kW of cooling). A dual-hose unit draws air to cool its condenser from outside instead of from inside the room, and thus is more effective than most single-hose units. These units create no negative pressure in the room.

Portable Evaporative System

Evaporative coolers, sometimes called "swamp coolers", do not have a compressor or condenser. Liquid water is evaporated on the cooling fins, releasing the vapor into the cooled area. Evaporating water absorbs a significant amount of heat, the latent heat of vaporisation, cooling the air. Humans and animals use the same mechanism to cool themselves by sweating.

Evaporative coolers have the advantage of needing no hoses to vent heat outside the cooled area, making them truly portable. They are also very cheap to install and use less energy than refrigerative air conditioners.

Uses

Air-conditioning engineers broadly divide air conditioning applications into *comfort* and *process* applications.

Comfort Applications

An array of air conditioners outside a commercial office building.

Comfort applications aim to provide a building indoor environment that remains relatively constant despite changes in external weather conditions or in internal heat loads.

Air conditioning makes deep plan buildings feasible, for otherwise they would have to be built narrower or with light wells so that inner spaces received sufficient outdoor air via natural ventilation. Air conditioning also allows buildings to be taller, since wind speed increases significantly with altitude making natural ventilation impractical for very tall buildings. Comfort applications are quite different for various building types and may be categorized as:

- Commercial buildings, which are built for commerce, including offices, malls, shopping centers, restaurants, etc.

- High-rise residential buildings, such as tall dormitories and apartment blocks

- Industrial spaces where thermal comfort of workers is desired

- Institutional buildings, which includes government buildings, hospitals, schools, etc.

- Low-rise residential buildings, including single-family houses, duplexes, and small apartment buildings

- Sports stadiums, such as the University of Phoenix Stadium and in Qatar for the 2022 FIFA World Cup.

Women have, on average, a significantly lower resting metabolic rate than men. Using inaccurate metabolic rate guidelines for air conditioning sizing can result in oversized and less efficient equipment, and setting system operating setpoints too cold can result in reduced worker productivity.

In addition to buildings, air conditioning can be used for many types of transportation, including automobiles, buses and other land vehicles, trains, ships, aircraft, and spacecraft.

Domestic usage

Typical residential central air conditioner in North America

Air conditioning is common in the US, with 88% of new single-family homes constructed in 2011 including air conditioning, ranging from 99% in the South to 62% in the West. In Canada, air conditioning use varies by province. In 2013, 55% of

Canadian households reported having an air conditioner, with high use in Manitoba (80%), Ontario (78%), Saskatchewan (67%), and Quebec (54%) and lower use in Prince Edward Island (23%), British Columbia (21%), and Newfoundland and Labrador (9%). In Europe, home air conditioning is generally less common. Southern European countries such as Greece have seen a wide proliferation of home air-conditioning units in recent years. In another southern European country, Malta, it is estimated that around 55% of households have an air conditioner installed. In India AC sales have dropped by 40%   due to higher costs and stricter energy efficiency regulations.

Process Applications

Process applications aim to provide a suitable environment for a process being carried out, regardless of internal heat and humidity loads and external weather conditions. It is the needs of the process that determine conditions, not human preference. Process applications include these:

- Chemical and biological laboratories

- Cleanrooms for the production of integrated circuits, pharmaceuticals, and the like, in which very high levels of air cleanliness and control of temperature and humidity are required for the success of the process.

- Environmental control of data centers

- Facilities for breeding laboratory animals. Since many animals normally reproduce only in spring, holding them in rooms in which conditions mirror those of spring all year can cause them to reproduce year-round.

- Food cooking and processing areas

- Hospital operating theatres, in which air is filtered to high levels to reduce infection risk and the humidity controlled to limit patient dehydration. Although temperatures are often in the comfort range, some specialist procedures, such as open heart surgery, require low temperatures (about 18 °C, 64 °F) and others, such as neonatal, relatively high temperatures (about 28 °C, 82 °F).

- Industrial environments

- Mining

- Nuclear power facilities

- Physical testing facilities

- Plants and farm growing areas

- Textile manufacturing

In both comfort and process applications, the objective may be to not only control temperature, but also humidity, air quality, and air movement from space to space.

Health Effects

Air-conditioning systems can promote the growth and spread of microorganisms, such as *Legionella pneumophila*, the infectious agent responsible for Legionnaires' disease, or thermophilic actinomycetes; however, this is only prevalent in poorly maintained water cooling towers. As long as the cooling tower is kept clean (usually by means of a chlorine treatment), these health hazards can be avoided or reduced.

Conversely, air conditioning (including filtration, humidification, cooling and disinfection) can be used to provide a clean, safe, hypoallergenic atmosphere in hospital operating rooms and other environments where proper atmosphere is critical to patient safety and well-being. Excessive air conditioning can have a negative effect on skin, causing it to dry out, and can also cause dehydration.

Environmental Impacts

Power Consumption

Innovation in air conditioning technologies continues, with much recent emphasis placed on energy efficiency. Production of the electricity used to operate air conditioners has an environmental impact, including the release of greenhouse gases.

Cylinder unloaders are a method of load control used mainly in commercial air conditioning systems. On a semi-hermetic (or open) compressor, the heads can be fitted with unloaders which remove a portion of the load from the compressor so that it can run better when full cooling is not needed. Unloaders can be electrical or mechanical.

In an automobile, the A/C system will use around 4 horsepower (3 kW) of the engine's power, thus increasing fuel consumption of the vehicle.

Refrigerants

Most refrigerants used for air conditioning contribute to global warming, and many also deplete the ozone layer. CFCs, HCFCs, and HFCs are potent greenhouse gases when leaked to the atmosphere.

The use of CFC as a refrigerant was once common, being used in the refrigerants R-11 and R-12 (sold under the brand name *Freon-12*). Freon refrigerants were commonly used during the 20th century in air conditioners due to their superior stability and safety properties. However, these chlorine-bearing refrigerants reach the upper atmosphere when they escape. Once the refrigerant reaches the stratosphere, UV radiation from the Sun homolytically cleaves the chlorine-carbon bond, yielding a chlorine radical. These chlorine radicals catalyze the breakdown of ozone into diatomic oxygen, depleting the ozone layer that shields the Earth's surface from strong UV radiation. Each chlorine radical remains active as a catalyst until it binds with another radical, forming a stable molecule and breaking the chain reaction.

Prior to 1994, most automotive air conditioning systems used R-12 as a refrigerant. It was replaced with R-134a refrigerant, which has no ozone depletion potential. Old R-12 systems can be retrofitted to R-134a by a complete flush and filter/dryer replacement to remove the mineral oil, which is not compatible with R-134a.

R22 (also known as HCFC-22) has a global warming potential about 1,800 times higher than CO_2. It was phased out for use in new equipment by 2010, and is to be completely discontinued by 2020. Although these gasses can be recycled when air conditioning units are disposed of, uncontrolled dumping and leaking can release gas directly into the atmosphere.

In the UK, the Ozone Regulations came into force in 2000 and banned the use of ozone depleting HCFC refrigerants such as R22 in new systems. The Regulation banned the use of R22 as a "top-up" fluid for maintenance between 2010 (for virgin fluid) and 2015 (for recycled fluid). This means that equipment that uses R22 can still operate, as long as it does not leak. Although R22 is now banned, units that use the refrigerant can still be serviced and maintained.

The manufacture and use of CFCs has been banned or severely restricted due to concerns about ozone depletion. In light of these environmental concerns, beginning on November 14, 1994, the U.S. Environmental Protection Agency has restricted the sale, possession and use of refrigerant to only licensed technicians, per rules under sections 608 and 609 of the Clean Air Act.

As an alternative to conventional refrigerants, other gases, such as CO_2 (R-744), have been proposed. R-744 is being adopted as a refrigerant in Europe and Japan. It is an effective refrigerant with a global warming potential of 1, but it must use higher compression to produce an equivalent cooling effect.

In 1992, a non-governmental organization, Greenpeace, was spurred by corporate executive policies and requested that a European lab find substitute refrigerants. This led to two alternatives, one a blend of propane (R290) and isobutane (R600a), and one of pure isobutane. Industry resisted change in Europe until 1993, and in the U.S. until 2011, despite some supportive steps in 2004 and 2008.

Ventilation (Architecture)

Ventilation is the intentional introduction of outside air into a space. Ventilation is mainly used to control indoor air quality by diluting and displacing indoor pollutants; it can also be used for purposes of thermal comfort or dehumidification when the introduction of outside air will help to achieve desired indoor psychrometric conditions.

An ab anbar (water reservoir) with double domes and windcatchers (openings near the top of the towers) in the central desert city of Naeen, Iran. Windcatchers are a form of natural ventilation.

The intentional introduction of outside air can be categorized as either mechanical ventilation, or natural ventilation. Mechanical ventilation uses fans to drive the flow of outside air into a building. This may be accomplished by pressurization (in the case of positively pressurized buildings), or by depressurization (in the case of exhaust ventilation systems). Many mechanically ventilated buildings use a combination of both, with the ventilation being integrated into the HVAC system. Natural ventilation is the intentional passive flow of outside air into a building through planned openings (such as louvers, doors, and windows). Natural ventilation does not require mechanical systems to move outside air, it relies entirely on passive physical phenomena, such as diffusion, wind pressure, or the stack effect. Mixed mode ventilation systems use both mechanical and natural processes. The mechanical and natural components may be used in conjunction with each other or separately at different times of day or season of the year. Since the natural component can be affected by unpredictable environmental conditions it may not always provide an appropriate amount of ventilation. In this case, mechanical systems may be used to supplement or to regulate the naturally driven flow.

Outdoor air can also enter a building by infiltration - the uncontrolled flow of air from outdoors to indoors through leaks (unplanned openings) in a building envelope. In buildings that make no intentional design for mechanical or natural ventilation, circumstantial infiltration has been referred to as adventitious ventilation. In exhaust ventilated buildings, the intended flow of outside air may enter through planned inlets, but it will also enter through unplanned leaks in the building envelope. Generally, all outside air that crosses the building envelope through leaks is referred to as infiltration, whether it is driven by mechanical systems or natural mechanisms like wind.

In many instances, ventilation for indoor air quality is simultaneously beneficial for the control of thermal comfort. At these times, it can be useful to increase the rate of ventilation beyond the minimum required for indoor air quality. Two examples include air-side economizer strategies and ventilation pre-cooling. In other instances, ventilation for indoor air quality contributes to the need for - and energy use by - mechanical heating and cooling equipment. It hot and humid climates dehumidification of ventilation air can be a particularly energy intensive process.

In many scenarios, heat recovery ventilation can reduce energy use for heating and cooling by facilitating sensible heat exchange between exhaust air and incoming ventilation air. Energy recovery ventilation transfers moisture in addition to sensible heat. However, heat recovery can increase the fan power required for ventilation, and may increase energy use for heating and cooling for periods when ventilation would be beneficial for the control of indoor thermal comfort.

The design of buildings that promote occupant health and well being requires clear understanding of the ways that ventilation airflow interacts with, dilutes, displaces or introduces pollutants within the occupied space. Although ventilation is an integral component to maintaining good indoor air quality, it may not be satisfactory alone. In scenarios where outdoor pollution would deteriorate indoor air quality, other treatment devices such as filtration may also be necessary. In kitchen ventilation systems, or for laboratory fume hoods, the design of effective effluent capture can be more important than the bulk amount of ventilation in a space. More generally, the way that an air distribution system causes ventilation to flow into and out of a space impacts the ability for a particular ventilation rate to remove internally generated pollutants. The ability for a system to remove pollution is described as its "ventilation effectiveness". However, the overall impacts of ventilation on indoor air quality can depend on more complex factors such as the sources of pollution, and the ways that activities and airflow interact to affect occupant exposure.

Ventilation should not be confused with air motion induced by ceiling fans or other devices. Air motion influences thermal comfort, it can decrease thermal stratification, and it may cause pollutant dilution by way of mixing, but it does not introduce outside air and therefore does not classify as ventilation.

Ventilation should be considered for its relationship to "venting" for appliances and combustion equipment such as water heaters, furnaces, boilers, and wood stoves. Most importantly, the design of building ventilation must be careful to avoid the backdraft of combustion products from "naturally vented" appliances into the occupied space. This issue is of greater importance in new buildings with more air tight envelopes. To avoid the hazard, many modern combustion appliances utilize "direct venting" which draws combustion air directly from outdoors, instead of from the indoor environment.

Categories of Ventilation

- Mechanical ventilation refers to any system that uses mechanical means, such as a fan, to introduce outside air to a space. This includes positive pressure ventilation, exhaust ventilation, and balanced systems that use both supply and exhaust ventilation.

- Natural ventilation refers to intentionally designed passive methods of introducing outside to a space without the use of mechanical systems.

- Mixed mode ventilation (or hybrid ventilation) systems use both natural and mechanical processes.

- Infiltration is the uncontrolled flow of air from outdoors to indoors through leaks (unplanned openings) in a building envelope. When a building design relies on environmentally driven circumstantial infiltration to maintain indoor air quality, this flow has been referred to as adventitious ventilation.

Ventilation Rate Standards

The ventilation rate, for CII buildings, is normally expressed by the volumetric flowrate of outside air being introduced to the building. The typical units used are cubic feet per minute (CFM) or liters per second (L/s). The ventilation rate can also be expressed on a per person or per unit floor area basis, such as CFM/p or CFM/ft², or as *air changes per hour* (ACH).

Standards for Residential Buildings

For residential buildings, which mostly rely on infiltration for meeting their ventilation needs, a common ventilation rate measure is the air change rate (or air changes per hour): the hourly ventilation rate divided by the volume of the space (*I* or *ACH*; units of 1/h). During the winter, ACH may range from 0.50 to 0.41 in a tightly insulated house to 1.11 to 1.47 in a loosely insulated house.

ASHRAE now recommends ventilation rates dependent upon floor area, as a revision to the 62-2001 standard, in which the minimum ACH was 0.35, but no less than 15 CFM/person (7.1 L/s/person). As of 2003, the standard has been changed to 3 CFM/100 sq. ft. (15 l/s/100 sq. m.) plus 7.5 CFM/person (3.5 L/s/person).

Standards for Commercial Buildings

Ventilation Rate Procedure

Ventilation Rate Procedure is rate based on standard and prescribes the rate at which ventilation air must be delivered to a space and various means to condition that air. Air quality is assessed (through CO_2 measurement) and ventilation rates are mathematically derived using constants. Indoor Air Quality Procedure uses one or more guidelines for the specification of acceptable concentrations of certain contaminants in indoor air but does not prescribe ventilation rates or air treatment methods. This addresses both quantitative and subjective evaluations, and is based on the Ventilation Rate Procedure. It also accounts for potential contaminants that may have no measured limits, or for which no limits are not set (such as formaldehyde offgassing from carpet and furniture).

History and Development of Ventilation rate Standards

- In 1973, in response to the 1973 oil crisis and conservation concerns, ASHRAE Standards 62-73 and 62-81) reduced required ventilation from 10 CFM (4.76 L/S) per person to 5 CFM (2.37 L/S) per person. This was found to be a primary cause of sick building syndrome.

- The 1989 ASHRAE standard (Standard 62-89) states that appropriate ventilation guidelines are 20 CFM (9.2 L/s) per person in an office building, and 15 CFM (7.1 L/s) per person for schools, while the 2004 Standard 62.1-2004 has lower recommendations again.

In certain applications, such as submarines, pressurized aircraft, and spacecraft, ventilation air is also needed to provide oxygen, and to dilute carbon dioxide for survival. Batteries in submarines also discharge hydrogen gas, which must also be ventilated for health and safety. In any pressurized, regulated environment, ventilation is necessary to control any fires that may occur, as the flames may be deprived of oxygen.

ANSI/ASHRAE (Standard 62-89) speculated that "comfort (odor) criteria are likely to be satisfied if the ventilation rate is set so that 1,000 ppm CO_2 is not exceeded" while OSHA has set a limit of 5000 ppm over 8 hours.

Ventilation guidelines are based upon the minimum ventilation rate required to maintain acceptable levels of bioeffluents. Carbon dioxide is used as a reference point, as it is the gas of highest emission at a relatively constant value of 0.005 L/s. The mass balance equation is:

$Q = G/(C_i - C_a)$

- Q = ventilation rate (L/s)

- $G = CO_2$ generation rate

- C_i = acceptable indoor CO_2 concentration

- C_a = ambient CO_2 concentration

Ventilating a space with fresh air aims to avoid "bad air". The study of what constitutes bad air dates back to the 1600s, when the scientist Mayow studied asphyxia of animals in confined bottles. The poisonous component of air was later identified as carbon dioxide (CO2), by Lavoisier in the very late 1700s, starting a debate as to the nature of "bad air" which humans perceive to be stuffy or unpleasant. Early hypotheses included excess concentrations of CO2 and oxygen depletion. However, by the late 1800s, scientists thought biological contamination, not oxygen or CO2, as the primary component of unacceptable indoor air. However, it was noted as early as 1872 that CO2 concentration closely correlates to perceived air quality.

The first estimate of minimum ventilation rates was developed by Tredgold in 1836. This was followed by subsequent studies on the topic by Billings in 1886 and Flugge in 1905. The recommendations of Billings and Flugge were incorporated into numerous building codes from 1900-1920s, and published as an industry standard by ASHVE (the predecessor to ASHRAE) in 1914.

Study continued into the varied effects of thermal comfort, oxygen, carbon dioxide, and biological contaminants. Research was conducted with humans subjects controlled test chambers. Two studies, published between 1909 and 1911, showed that carbon dioxide was not the offending component. Subjects remained satisfied in chambers with high levels of CO2, so long as the chamber remained cool. (Subsequently, it has been determined that CO2 is, in fact, harmful at concentrations over 50,000ppm)

ASHVE began a robust research effort in 1919. By 1935, ASHVE funded research conducted by Lemberg, Brandt, and Morse - again using human subjects in test chambers - suggested the primary component of "bad air" was odor, perceived by the human olfactory nerves. Human response to odor was found to be logarithmic to contaminant concentrations, and related to temperature. At lower, more comfortable temperatures, lower ventilation rates were satisfactory. A 1936 human test chamber study by Yaglou, Riley, and Coggins culminated much of this effort, considering odor, room volume, occupant age, cooling equipment effects, and recirculated air implications, which provided guidance for ventilation rates. The Yaglou research has been validated, and adopted into industry standards, beginning with the ASA code in 1946. From this research base, ASHRAE (having replaced ASHVE) developed space by space recommendations, and published them as ASHRAE Standard 62-1975: Ventilation for acceptable indoor air quality.

As more architecture incorporated mechanical ventilation, the cost of outdoor air ventilation came under some scrutiny. In cold, warm, humid, or dusty climates, it is preferable to minimize ventilation with outdoor air to conserve energy, cost, or filtration. This critique (e.g. Tiller) led ASHRAE to reduce outdoor ventilation rates in 1981, particularly in non-smoking areas. However subsequent research by Fanger, W. Cain, and Janssen validated the Yaglou model.

Historical Ventilation Rates				
Author or Source	Year	Ventilation Rate (IP)	Ventilation Rate (SI)	Basis or rationale
Tredgold	1836	4 CFM per person	2 L/s per person	Basic metabolic needs, breathing rate, and candle burning
Billings	1895	30 CFM per person	15 L/s per person	Indoor air hygiene, preventing spread of disease

Flugge	1905	30 CFM per person	15 L/s per person	Excessive temperature or unpleasant odor
ASHVE	1914	30 CFM per person	15 L/s per person	Based on Billings, Flugge and contemporaries
Early US Codes	1925	30 CFM per person	15 L/s per person	Same as above
Yaglou	1936	15 CFM per person	7.5 L/s per person	Odor control, outdoor air as a fraction of total air
ASA	1946	15 CFM per person	7.5 L/s per person	Based on Yahlou and contemporaries
ASHRAE	1975	15 CFM per person	7.5 L/s per person	Same as above
ASHRAE	1981	10 CFM per person	5 L/s per person	For non-smoking areas, reduced.
ASHRAE	1989	15 CFM per person	7.5 L/s per person	Based on Fanger, W. Cain, and Janssen

ASHRAE continues to publish space-by-space ventilation rate recommendations, which are decided by a consensus committee of industry experts. The modern descendants of ASHRAE standard 62-1975 are ASHRAE Standard 62.1, for non-residential spaces, and ASHRAE 62.2 for residences.

In 2004, the calculation method was revised to include both an occupant-based contamination component and an area–based contamination component. These two components are additive, to arrive at an overall ventilation rate. The change was made to recognize that densely populated areas were sometimes overventilated (leading to higher energy and cost) using a per-person methodology.

Occupant Based Ventilation Rates, ANSI/ASHRAE Standard 62.1-2004

IP Units	SI Units	Category	Examples
0 cfm/ person	0 L/s/ person	Spaces where ventilation requirements are primarily associated with building elements, not occupants.	Storage Rooms, Warehouses
5 cfm/person	2.5 L/s/ person	Spaces occupied by adults, engaged in low levels of activity	Office space
7.5 cfm/ person	3.5 L/s/ person	Spaces where occupants are engaged in higher levels of activity, but not strenuous, or activities generating more contaminants	Retail spaces, lobbies
10 cfm/ person	5 L/s/ person	Spaces where occupants are engaged in more strenuous activity, but not exercise, or activities generating more contaminants	Classrooms, school settings
20 cfm/ person	10 L/s/ person	Spaces where occupants are engaged in exercise, or activities generating many contaminants	dance floors, exercise rooms

Area-based ventilation rates, ANSI/ASHRAE Standard 62.1-2004

IP Units	SI Units	Category	Examples
0.06 cfm/ft²	0.30 L/s/m²	Spaces where space contamination is normal, or similar to an office environment	Conference rooms, lobbies
0.12 cfm/ft²	0.60 L/s/m²	Spaces where space contamination is significantly higher than an office environment	Classrooms, museums
0.18 cfm/ft²	0.90 L/s/m²	Spaces where space contamination is even higher than the previous category	Laboratories, art classrooms
0.30 cfm/ft²	1.5 L/s/m²	Specific spaces in sports or entertainment where contaminants are released	Sports, entertainment
0.48 cfm/ft²	2.4 L/s/m²	Reserved for indoor swimming areas, where chemical concentrations are high	Indoor swimming areas

The addition of occupant- and area-based ventilation rates found in the tables above often results in significantly reduced rates compared to the former standard. This is compensated in other sections of the standard which require that this minimum amount of air is actually delivered to the breathing zone of the individual occupant at all times. The total outdoor air intake of the ventilation system (in multiple-zone variable air volume (VAV) systems) might therefore be similar to the airflow required by the 1989 standard. From 1999 to 2010, there was considerable development of the application protocol for ventilation rates. These advancements address occupant- and process-based ventilation rates, room ventilation effectiveness, and system ventilation effectiveness

Natural Ventilation Strategies

Techniques and architectural features used to ventilate buildings and structures naturally include, but are not limited to:

- Operable windows

- Pressurised air pumps

- Night purge ventilation

- Clerestory windows and vented skylights

- Building orientation

- Wind capture façades

Natural Ventilation

Natural ventilation harnesses naturally available forces to supply and remove air in an enclosed space. There are three types of natural ventilation occurring in buildings: wind driven ventilation, pressure-driven flows, and stack ventilation. The pressures generated by 'the stack effect' rely upon the buoyancy of heated or rising air. Wind driven ventilation relies upon the force of the prevailing wind to pull and push air through the enclosed space as well as through breaches in the building's envelope. Seoul University Professor Wonjun Kwon recently discovered a new way to ventilate large area of indoor space. The so-called "air pump" system uses pressure between inside and outside of rooms to push air out of a structure.

Almost all historic buildings were ventilated naturally. The technique was generally abandoned in larger US buildings during the late 20th century as the use of air conditioning became more widespread. However, with the advent of advanced Building Energy Modeling (BEM) software, improved Building Automation Systems (BAS), Leadership in Energy and Environmental Design (LEED) design requirements, and improved window manufacturing techniques; natural ventilation has made a resurgence in commercial buildings both globally and throughout the US.

The benefits of natural ventilation include:

- Improved Indoor air quality (IAQ)

- Energy savings

- Reduction of greenhouse gas emissions

- Occupant control

- Reduction in occupant illness associated with Sick Building Syndrome

- Increased worker productivity

Mechanical Ventilation Strategies

Mechanical ventilation of buildings and structures can be achieved by use of the following techniques:

- Whole-house ventilation

- Mixing ventilation

- Displacement ventilation

- Dedicated outside air supply

Demand-controlled Ventilation (DCV)

Demand-controlled ventilation (DCV, also known as Demand Control Ventilation) makes it possible to maintain air quality while conserving energy. ASHRAE has determined that: *"It is consistent with the ventilation rate procedure that demand control be permitted for use to reduce the total outdoor air supply during periods of less occupancy.*" In a DCV system, CO_2 sensors control the amount of ventilation. During peak occupancy, CO_2 levels rise, and the system adjusts to deliver the same amount of outdoor air as would be used by the ventilation-rate procedure. However, when spaces are less occupied, CO_2 levels reduce, and the system reduces ventilation to conserves energy. DCV is a well-established practice, and is required in high occupancy spaces by building energy standards such as ASHRAE 90.1.

Personalized Ventilation

Personalized ventilation is an air distribution strategy that allows individuals to control the amount of ventilation received. The approach deliver fresh air more directly to the breathing zone and aims to improve air quality of inhaled air. Personalized ventilation provides a much higher ventilation effectiveness than conventional mixing ventilation systems by displacing pollution from the breathing zone far less air volume. Beyond improved air quality benefits, the strategy can also improve occupant's thermal comfort, perceived air quality, and overall satisfaction with the indoor environment. Individual's preferences for temperature and air movement are not equal, and so traditional approaches to homogeneous environmental control have failed to achieve high occupant satisfaction. Techniques such as personalized ventilation facilitate control of a more diverse thermal environment that can improve thermal satisfaction for most occupants.

Local Exhaust Ventilation

Local exhaust ventilation addresses the issue of avoiding the contamination of indoor air by specific high-emission sources by capturing airborne contaminants before they are spread into the environment. This can include water vapor control, lavatory bioeffluent control, solvent vapors from industrial processes, and dust from wood- and metal-working machinery. Air can be exhausted through pressurized hoods or through the use of fans and pressurizing a specific area. A local exhaust system is composed of 5 basic parts

1. A hood that captures the contaminant at its source

2. Ducts for transporting the air

3. An air-cleaning device that removes/minimizes the contaminant

4. A fan that moves the air through the system

5. An exhaust stack through which the contaminated air is discharged

In the UK, the use of LEV systems have regulations set out by the Health and Safety Executive (HSE) which are referred to as the Control of Substances Hazardous to Health (CoSHH). Under CoSHH, legislation is set out to protect users of LEV systems by ensuring that all equipment is tested at least every fourteen months to ensure the LEV systems are performing adequately. All parts of the system must be visually inspected and thoroughly tested and where any parts are found to be defective the inspector must issue a red label to identify the defective part and the issue.

The owner of the LEV system must then have the defective parts repaired or replaced before the system can be used.

Ventilation and Combustion

Combustion (e.g., fireplace, gas heater, candle, oil lamp, etc.) consumes oxygen while producing carbon dioxide and other unhealthy gases and smoke, requiring ventilation air. An open chimney promotes infiltration (i.e. natural ventilation) because of the negative pressure change induced by the buoyant, warmer air leaving through the chimney. The warm air is typically replaced by heavier, cold air.

Ventilation in a structure is also needed for removing water vapor produced by respiration, burning, and cooking, and for removing odors. If water vapor is permitted to accumulate, it may damage the structure, insulation, or finishes. When operating, an air conditioner usually removes excess moisture from the air. A dehumidifier may also be appropriate for removing airborne moisture.

Smoking and Ventilation

ASHRAE standard 62 states that air removed from an area with environmental tobacco smoke shall not be recirculated into ETS-free air. A space with ETS requires more ventilation to achieve similar perceived air quality to that of a non-smoking environment.

The amount of ventilation in an ETS area is equal to the amount of ETS-free area plus the amount V, where:

$V = DSD \times VA \times A/60E$

- V = recommended extra flow rate in CFM (L/s)

- DSD = design smoking density (estimated number of cigarettes smoked per hour per unit area)

- VA = volume of ventilation air per cigarette for the room being designed (ft^3/cig)

- E = contaminant removal effectiveness

History

The development of forced ventilation was spurred by the common belief in the late 18th and early 19th century in the miasma theory of disease, where stagnant 'airs' were thought to spread illness. An early method of ventilation was the use of a ventilating fire near an air vent which would forcibly cause the air in the building to circulate. English engineer John Theophilus Desaguliers provided an early example of this, when he installed ventilating fires in the air tubes on the roof of the House of Commons. Starting with the Covent Garden Theatre, gas burning chandeliers on the ceiling were often specially designed to perform a ventilating role.

The Central Tower of the Palace of Westminster. This octagonal spire was for ventilation purposes, in the more complex system imposed by Reid on Barry, in which it was to draw air out of the Palace. The design was for aesthetic disguise of its function.

Mechanical Systems

A more sophisticated system involving the use of mechanical equipment to circulate the air was developed in the mid 19th century. A basic system of bellows was put in place to ventilate Newgate Prison and outlying buildings, by the engineer Stephen Hales in the mid-18th century. The problem with these early devices was that they required constant human labour to operate. David Boswell Reid was called to testify before a Parliamentary committee on proposed architectural designs for the new House of Commons, after the old one burned down in a fire in 1834. In January 1840 Reid was appointed by the committee for the House of Lords dealing with the construction of the replacement for the Houses of Parliament. The post was in the capacity of ventilation engineer, in effect; and with its creation there began a long series of quarrels between Reid and Charles Barry, the architect.

He advocated the installation of a very advanced ventilation system in the new House. His design had air being drawn into an underground chamber, where it would undergo either heating or cooling. It would then ascend into the chamber through thousands of small holes drilled into the floor, and would be extracted through the ceiling by a special ventilation fire within a great stack.

Reid's reputation was made by his work in Westminster. He was commissioned for an air quality survey in 1837 by the Leeds and Selby Railway in their tunnel. The steam vessels built for the Niger expedition of 1841 were fitted with ventilation systems based on Reid's Westminster model. Air was dried, filtered and passed over charcoal. Reid's ventilation method was also applied more fully to St. George's Hall, Liverpool, where the architect, Harvey Lonsdale Elmes, requested that Reid should be involved in ventilation design. Reid considered this the only building in which his system was completely carried out.

Fans

With the advent of practical steam power, fans could finally be used for ventilation. Reid installed four steam powered fans in the ceiling of St George's Hospital in Liverpool, so that the pressure produced by the fans would force the incoming air upward and through vents in the ceiling. Reid's pioneering work provides the basis for ventilation systems to this day. He was remembered as "Dr. Reid the ventilator" in the twenty-first century in discussions of energy efficiency, by Lord Wade of Chorlton.

Problems

- In hot, humid climates, unconditioned ventilation air will deliver approximately one pound of water each day for each cfm of outdoor air per day, annual average. This is a great deal of moisture, and it can create serious indoor moisture and mold problems.

- Ventilation efficiency is determined by design and layout, and is dependent upon placement and proximity of diffusers and return air outlets. If they are located closely together, supply air may mix with stale air, decreasing efficiency of the HVAC system, and creating air quality problems.

- System imbalances occur when components of the HVAC system are improperly adjusted or installed, and can create pressure differences (too much circulating air creating a draft or too little circulating air creating stagnancy).

- Cross-contamination occurs when pressure differences arise, forcing potentially contaminated air from one zone to an uncontaminated zone. This often involves undesired odors or VOCs.

- Re-entry of exhaust air occurs when exhaust outlets and fresh air intakes are either too close, or prevailing winds change exhaust patterns, or by infiltration between intake and exhaust air flows.

- Entrainment of contaminated outside air through intake flows will result in indoor air contamination. There are a variety of contaminated air sources, ranging from industrial effluent to VOCs put off by nearby construction work.

Air Handler

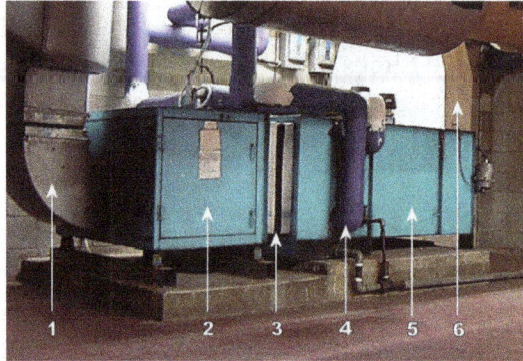

An air handling unit; air flow is from the right to left in this case. Some AHU components shown are 1 – Supply duct 2 – Fan compartment 3 – Vibration isolator ('flex joint') 4 – Heating and/or cooling coil 5 – Filter compartment 6 – Mixed (recirculated + outside) air duct

A rooftop packaged unit or RTU

An air handler, or air handling unit (often abbreviated to AHU), is a device used to regulate and circulate air as part of a heating, ventilating, and air-conditioning (HVAC) system. An air handler is usually a large metal box containing a blower, heating or cooling elements, filter racks or chambers, sound attenuators, and dampers. Air handlers usually connect to a ductwork ventilation system that distributes the conditioned air through the building and returns it to the AHU. Sometimes AHUs discharge (*supply*) and admit (*return*) air directly to and from the space served without ductwork.

Small air handlers, for local use, are called terminal units, and may only include an air filter, coil, and blower; these simple terminal units are called blower coils or fan coil units. A larger air handler that conditions 100% outside air, and no recirculated air, is known as a makeup air unit (MAU). An air handler designed for outdoor use, typically on roofs, is known as a packaged unit (PU) or rooftop unit (RTU).

Construction

The air handler is normally constructed around a framing system with metal infill panels as required to suit the configuration of the components. In its simplest form the frame may be made from metal channels or sections, with single skin metal infill panels. The metalwork is normally galvanized for long term protection. For outdoor units some form of weatherproof lid and additional sealing around joints is provided.

Larger air handlers will be manufactured from a square section steel framing system with double skinned and insulated infill panels. Such constructions reduce heat loss or heat gain from the air handler, as well as providing acoustic attenuation. Larger air handlers may be several meters long and are manufactured in a *sectional* manner and therefore, for strength and rigidity, steel section base rails are provided under the unit.

Where supply and extract air is required in equal proportions for a balanced ventilation system, it is common for the supply and extract air handlers to be joined together, either in a *side-by-side* or a *stacked* configuration.

Components

The major types of components are described here in approximate order, from the return duct (input to the AHU), through the unit, to the supply duct (AHU output).

Filters

Air filtration is almost always present in order to provide clean dust-free air to the building occupants. It may be via simple low-MERV pleated media, HEPA, electrostatic, or a combination of techniques. Gas-phase and ultraviolet air treatments may be employed as well.

Filtration is typically placed first in the AHU in order to keep all the downstream components clean. Depending upon the grade of filtration required, typically filters will be arranged in two (or more) successive banks with a coarse-grade panel filter provided in front of a fine-grade bag filter, or other "final" filtration medium. The panel filter is cheaper to replace and maintain, and thus protects the more expensive bag filters.

The life of a filter may be assessed by monitoring the pressure drop through the filter medium at design air volume flow rate. This may be done by means of a visual display using a pressure gauge, or by a pressure switch linked to an alarm point on the building control system. Failure to replace a filter may eventually lead to its collapse, as the forces exerted upon it by the fan overcome its inherent strength, resulting in collapse and thus contamination of the air handler and downstream ductwork.

Heating and/or Cooling Elements

Air handlers may need to provide heating, cooling, or both to change the supply air temperature, and humidity level depending on the location and the application. Such conditioning is provided by heat exchanger coil(s) within the air handling unit air stream, such coils may be *direct* or *indirect* in relation to the medium providing the heating or cooling effect.

Direct heat exchangers include those for gas-fired fuel-burning heaters or a refrigeration evaporator, placed directly in the air stream. Electric resistance heaters and heat pumps can be used as well. Evaporative cooling is possible in dry climates.

Indirect coils use hot water or steam for heating, and chilled water for cooling (prime energy for heating and cooling is provided by central plant elsewhere in the building). Coils are typically manufactured from copper for the tubes, with copper or aluminium fins to aid heat transfer. Cooling coils will also employ eliminator plates to remove and drain condensate. The hot water or steam is provided by a central boiler, and the chilled water is provided by a central chiller. Downstream temperature sensors are typically used to monitor and control "off coil" temperatures, in conjunction with an appropriate motorized control valve prior to the coil.

If dehumidification is required, then the cooling coil is employed to *over-cool* so that the dew point is reached and condensation occurs. A heater coil placed after the cooling coil re-heats the air (therefore known as a *re-heat coil*) to the desired supply temperature. This has the effect of reducing the relative humidity level of the supply air.

In colder climates, where winter temperatures regularly drop below freezing, then *frost coils* or *pre-heat* coils are often employed as a first stage of air treatment to ensure that downstream filters or chilled water coils are protected against freezing. The control of the frost coil is such that if a certain off-coil air temperature is not reached then the entire air handler is shut down for protection.

Humidifier

Humidification is often necessary in colder climates where continuous heating will make the air drier, resulting in uncomfortable air quality and increased static electricity. Various types of humidification may be used:

- Evaporative: dry air blown over a reservoir will evaporate some of the water. The rate of evaporation can be increased by spraying the water onto baffles in the air stream.

- Vaporizer: steam or vapor from a boiler is blown directly into the air stream.

- Spray mist: water is diffused either by a nozzle or other mechanical means into fine droplets and carried by the air.

- Ultrasonic: A tray of fresh water in the airstream is excited by an ultrasonic device forming a fog or water mist.

- Wetted medium: A fine fibrous medium in the airstream is kept moist with fresh water from a header pipe with a series of small outlets. As the air passes through the medium it entrains the water in fine droplets. This type of humidifier can quickly clog if the primary air filtration is not maintained in good order.

Mixing Chamber

In order to maintain indoor air quality, air handlers commonly have provisions to allow the introduction of outside air into, and the exhausting of air from the building. In temperate climates, mixing the right amount of cooler outside air with warmer return air can be used to approach the desired supply air temperature. A mixing chamber is therefore used which has dampers controlling the ratio between the return, outside, and exhaust air.

Blower/Fan

Air handlers typically employ a large squirrel cage blower driven by an AC induction electric motor to move the air. The blower may operate at a single speed, offer a variety of set speeds, or be driven by a Variable Frequency Drive to allow a wide range of air flow rates. Flow rate may also be controlled by inlet vanes or outlet dampers on the fan. Some residential air handlers in USA (central "furnaces" or "air conditioners") use a brushless DC electric motor that has variable speed capabilities. Air handlers in Europe and Australia and New Zealand now commonly use backward curve fans without scroll or "plug fans". These are driven using high efficiency EC (electronically commutated) motors with built in speed control.

Multiple blowers may be present in large commercial air handling units, typically placed at the end of the AHU and the beginning of the supply ductwork (therefore also called "supply fans"). They are often augmented by fans in the return air duct ("return fans") pushing the air into the AHU.

Balancing

Un-balanced fans wobble and vibrate. For home AC fans, this can be a major problem: air circulation is greatly reduced at the vents (as wobble is lost energy), efficiency is compromised, and noise is increased. Another major problem in fans that are not balanced is longevity of the bearings (attached to the fan and shaft) is compromised. This can cause failure to occur long before the bearings life expectancy.

Weights can be strategically placed to correct for a smooth spin (for a ceiling fan, trial and error placement typically resolves the problem). But for a home / central AC fan or big fan are typically taken to shops, which have special balancers for more complicated balancing (trial and error can cause damage before the correct points are found). The fan motor itself does not typically vibrate.

Heat Recovery Device

A heat recovery device heat exchanger of many types, may be fitted to the air handler between supply and extract airstreams for energy savings and increasing capacity. These types more commonly include for:

- Recuperator, or Plate Heat exchanger: A sandwich of plastic or metal plates with interlaced air paths. Heat is transferred between airstreams from one side of the plate to the other. The plates are typically spaced at 4 to 6mm apart. Can also be used to recover coolth. Heat recovery efficiency up to 70%.

- Thermal Wheel, or Rotary heat exchanger: A slowly rotating matrix of finely corrugated metal, operating in both opposing airstreams. When the air handling unit is in heating mode, heat is absorbed as air passes through the matrix in the exhaust airstream, during one half rotation, and released during the second half rotation into the supply airstream in a continuous process. When the air handling unit is in cooling mode, heat is released as air passes through the matrix in the exhaust airstream, during one half rotation, and absorbed during the second half rotation into the supply airstream. Heat recovery efficiency up to 85%. Wheels are also available with a hydroscopic coating to provide latent heat transfer and also the drying or humidification of airstreams.

- Run around coil: Two air to liquid heat exchanger coils, in opposing airstreams, piped together with a circulating pump and using water or a brine as the heat transfer medium. This device, although not very efficient, allows heat recovery between remote and sometimes multiple supply and exhaust airstreams. Heat recovery efficiency up to 50%.

- Heat Pipe: Operating in both opposing air paths, using a confined refrigerant as a heat transfer medium. The heat pipe uses multiple sealed pipes mounted in a coil configuration with fins to increase heat transfer. Heat is absorbed on one side of the pipe, by evaporation of the refrigerant, and released at the other side, by condensation of the refrigerant. Condensed refrigerant flows by gravity to the first side of the pipe to repeat the process. Heat recovery efficiency up to 65%.

Controls

Controls are necessary to regulate every aspect of an air handler, such as: flow rate of air, supply air temperature, mixed air temperature, humidity, air quality. They may be as simple as an off/on thermostat or as complex as a building automation system using BACnet or LonWorks, for example.

Common control components include temperature sensors, humidity sensors, sail switches, actuators, motors, and controllers.

Vibration Isolators

The blowers in an air handler can create substantial vibration and the large area of the duct system would transmit this noise and vibration to the occupants of the building. To avoid this, vibration isolators (flexible sections) are normally inserted into the duct immediately before and after the air handler and often also between the fan compartment and the rest of the AHU. The rubberized canvas-like material of these sections allows the air handler components to vibrate without transmitting this motion to the attached ducts.

The fan compartment can be further isolated by placing it on a spring suspension, which will mitigate the transfer of vibration through the floor.

References

- May, Jeffrey C. (2001). My house is killing me! : the home guide for families with allergies and asthma. Baltimore: The Johns Hopkins University Press. ISBN 978-0-8018-6730-9.

- Salthammer, T., ed. (1999). Organic Indoor Air Pollutants — Occurrence, Measurement, Evaluation. Wiley-VCH. ISBN 3-527-29622-0.

- Spengler, J.D.; Samet, J.M. (1991). Indoor air pollution: A health perspective. Baltimore: Johns Hopkins University Press. ISBN 0-8018-4125-9.

- Tichenor, B. (1996). Characterizing Sources of Indoor Air Pollution and Related Sink Effects. ASTM STP 1287. West Conshohocken, PA: ASTM. ISBN 0-8031-2030-3.

- Swenson, S. Don (1995). HVAC: heating, ventilating, and air conditioning. Homewood, Illinois: American Technical Publishers. ISBN 978-0-8269-0675-5.

- Al-Kodmany, Kheir (2013). The Future of the City: Tall Buildings and Urban Design. WIT Press. p. 242. ISBN 978-1-84564-410-9.

- Needham, Joseph (1991). Science and Civilisation in China, Volume 4: Physics and Physical Technology, Part 2, Mechanical Engineering. Cambridge University Press. pp. 99, 151, 233. ISBN 978-0-521-05803-2.

- Shane Smith (2000). Greenhouse gardener's companion: growing food and flowers in your greenhouse or sunspace (2nd ed.). Fulcrum Publishing. p. 62. ISBN 978-1-55591-450-9.

- Hussain, Ali Vendavarz, Sunil Kumar, Muhammed (2006). HVAC : handbook of heating, ventilation, and air conditioning (1st ed.). New York: Industrial Press. ISBN 0-8311-3163-2.

Air Pollution Control Technologies

Cyclonic separation is the method of removing particulates; these particulates are removed from air, gas and also from liquid streams. The alternative technologies used are selective catalytic reduction, exhaust gas recirculation, biofilter, thermal oxidizer and vapor recovery. This section serves as a source to understand the main technologies used in controlling air pollution.

Cyclonic Separation

Cyclonic separation is a method of removing particulates from an air, gas or liquid stream, without the use of filters, through vortex separation. When removing particulate matter from liquids, a hydrocyclone is used; while from gas, a gas cyclone is used. Rotational effects and gravity are used to separate mixtures of solids and fluids. The method can also be used to separate fine droplets of liquid from a gaseous stream.

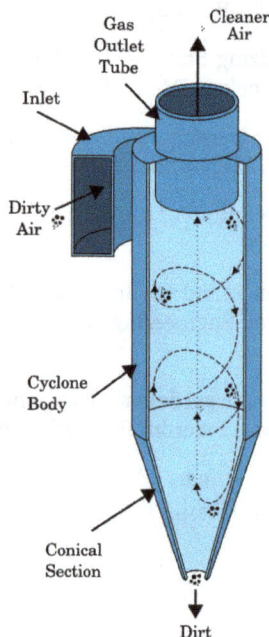

A simple cyclone separator

A high speed rotating (air)flow is established within a cylindrical or conical container called a cyclone. Air flows in a helical pattern, beginning at the top (wide end) of the cyclone and ending at the bottom (narrow) end before exiting the cyclone in a straight stream through the center of the cyclone and out the top. Larger (denser) particles in the rotating stream have too much inertia to follow the tight curve of the stream, and strike the outside wall, then fall to the bottom of the cyclone where they can be removed. In a conical system, as the rotating flow moves towards the narrow end of the cyclone, the rotational radius of the stream is reduced, thus separating smaller and smaller particles. The cyclone geometry, together with flow rate, defines the *cut point* of the cyclone. This is the size of particle that will be removed from the stream with a 50% efficiency. Particles larger than the cut point will be removed with a greater efficiency, and smaller particles with a lower efficiency.

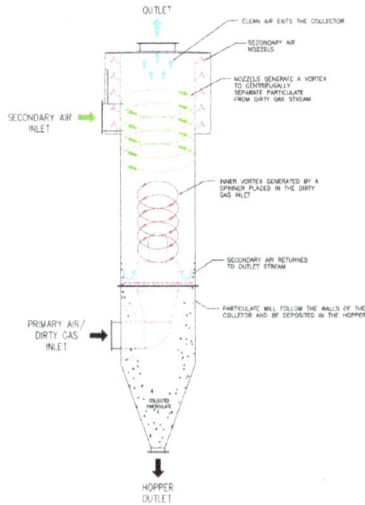

Airflow diagram for Aerodyne cyclone in standard vertical position.
Secondary air flow is injected to reduce wall abrasion.

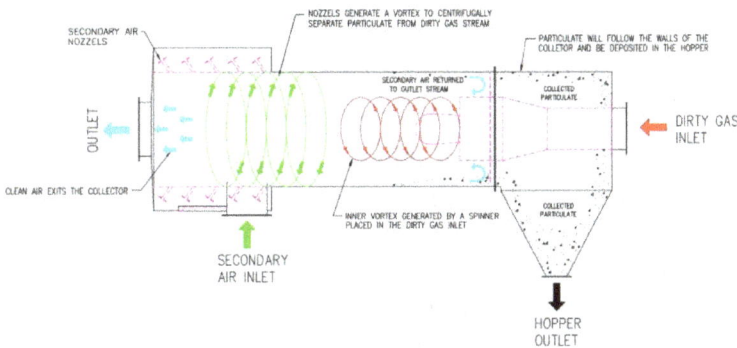

Airflow diagram for Aerodyne cyclone in horizontal position, an alternate design. Secondary air flow is injected to reduce wall abrasion, and to help move collected particulates to hopper for extraction.

An alternative cyclone design uses a secondary air flow within the cyclone to keep the collected particles from striking the walls, to protect them from abrasion. The primary air flow containing the particulates enters from the bottom of the cyclone and is forced into spiral rotation by stationary spinner vanes. The secondary air flow enters from the top of the cyclone and moves downward toward the bottom, intercepting the particulate from the primary air. The secondary air flow also allows the collector to optionally be mounted horizontally, because it pushes the particulate toward the collection area, and does not rely solely on gravity to perform this function.

Large scale cyclones are used in sawmills to remove sawdust from extracted air. Cyclones are also used in oil refineries to separate oils and gases, and in the cement industry as components of kiln preheaters. Cyclones are increasingly used in the household, as the core technology in bagless types of portable vacuum cleaners and central vacuum cleaners. Cyclones are also used in industrial and professional kitchen ventilation for separating the grease from the exhaust air in extraction hoods. Smaller cyclones are used to separate airborne particles for analysis. Some are small enough to be worn clipped to clothing, and are used to separate respirable particles for later analysis.

Similar separators are used in the oil refining industry (e.g. for Fluid catalytic cracking) to achieve fast separation of the catalyst particles from the reacting gases and vapors.

James Dyson has become a billionaire from developing and marketing bagless vacuum cleaners based on cyclonic separation of dust, initially inspired by seeing sawdust separator at a sawmill.

Analogous devices for separating particles or solids from liquids are called hydrocyclones or hydroclones. These may be used to separate solid waste from water in wastewater and sewage treatment.

Cyclone Theory

As the cyclone is essentially a two phase particle-fluid system, fluid mechanics and particle transport equations can be used to describe the behaviour of a cyclone. The air in a cyclone is initially introduced tangentially into the cyclone with an inlet velocity V_{in}. Assuming that the particle is spherical, a simple analysis to calculate critical separation particle sizes can be established.

If one considers an isolated particle circling in the upper cylindrical component of the cyclone at a rotational radius of r from the cyclone's central axis, the particle is therefore subjected to drag, centrifugal, and buoyant forces. Given that the fluid velocity is moving in a spiral the gas velocity can be broken into two component velocities: a tangential component, V_t, and an outward radial velocity component V_r. Assuming Stokes' law, the drag force in the outward radial direction that is opposing the outward velocity on any particle in the inlet stream is:

$$F_d = -6\pi r_p \mu V_r.$$

Using ρ_p as the particles density, the centrifugal component in the outward radial direction is:

$$F_c = m\frac{V_t^2}{r} = \frac{4}{3}\pi\rho_p r_p^3 \frac{V_t^2}{r}.$$

The buoyant force component is in the inward radial direction. It is in the opposite direction to the particle's centrifugal force because it is on a volume of fluid that is missing compared to the surrounding fluid. Using ρ_f for the density of the fluid, the buoyant force is:

$$F_b = -V_p \rho_f \frac{V_t^2}{r} = -\frac{4\pi r_p^3}{3}\frac{V_t^2}{r}\rho_f.$$

In this case, V_p is equal to the volume of the particle (as opposed to the velocity). Determining the outward radial motion of each particle is found by setting Newton's second law of motion equal to the sum of these forces:

$$m\frac{dV_r}{dt} = F_d + F_c + F_b$$

To simplify this, we can assume the particle under consideration has reached "terminal velocity", i.e., that its acceleration $\frac{dV_r}{dt}$ is zero. This occurs when the radial velocity has caused enough drag force to counter the centrifugal and buoyancy forces. This simplification changes our equation to:

$$F_d + F_c + F_b = 0$$

Which expands to:

$$-6\pi r_p \mu V_r + \frac{4}{3}\pi r_p^3 \frac{V_t^2}{r}\rho_p - \frac{4}{3}\pi r_p^3 \frac{V_t^2}{r}\rho_f = 0$$

Solving for V_r we have

$$V_r = \frac{2}{9}\frac{r_p^2}{\mu}\frac{V_t^2}{r}(\rho_p - \rho_f)$$

Notice that if the density of the fluid is greater than the density of the particle, the motion is (-), toward the center of rotation and if the particle is denser than the fluid, the motion is (+), away from the center. In most cases, this solution is used as guidance in designing a separator, while actual performance is evaluated and modified empirically.

In non-equilibrium conditions when radial acceleration is not zero, the general equation from above must be solved. Rearranging terms we obtain

$$\frac{dV_r}{dt} + \frac{9}{2}\frac{\mu}{\rho_p r_p^2}V_r - \left(1 - \frac{\rho_f}{\rho_p}\right)\frac{V_t^2}{r} = 0$$

Since V_r is distance per time, this is a 2nd order differential equation of the form

$$x'' + c_1 x' + c_2 = 0..$$

Experimentally it is found that the velocity component of rotational flow is proportional to r^2, therefore:

$$V_t \propto r^2.$$

This means that the established feed velocity controls the vortex rate inside the cyclone, and the velocity at an arbitrary radius is therefore:

$$U_r = U_{in}\frac{r}{R_{in}}.$$

Subsequently, given a value for V_t, possibly based upon the injection angle, and a cutoff radius, a characteristic particle filtering radius can be estimated, above which particles will be removed from the gas stream.

Alternative Models

The above equations are limited in many regards. For example, the geometry of the separator is not considered, the particles are assumed to achieve a steady state and the effect of the vortex inversion at the base of the cyclone is also ignored, all behaviours which are unlikely to be achieved in a cyclone at real operating conditions.

More complete models exist, as many authors have studied the behaviour of cyclone separators. Numerical modelling using computational fluid dynamics has also been used extensively in the study of cyclonic behaviour. A major limitation of any fluid mechanics model for cyclone separators is the inability to predict the agglomeration of fine particles with larger particles, which has a great impact on cyclone collection efficiency.

Selective Catalytic Reduction

Selective catalytic reduction (SCR) is a means of converting nitrogen oxides, also referred to as NO x with the aid of a catalyst into diatomic nitrogen (N2) , and water (H2O). A gaseous reductant, typically anhydrous ammonia, aqueous ammonia or urea, is added to a stream of flue or exhaust gas and is adsorbed onto a catalyst. Carbon dioxide, CO2 is a reaction product when urea is used as the reductant.

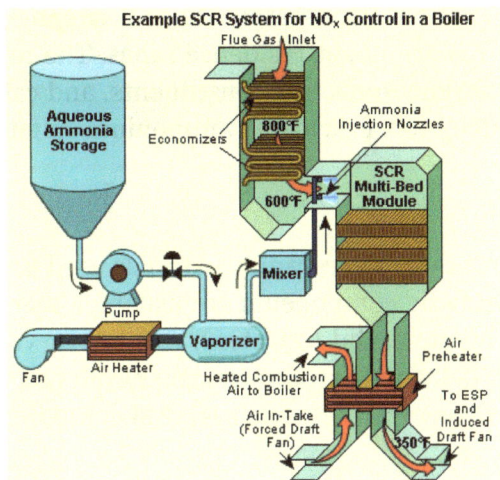

An aqueous ammonia SCR process overview; a vaporizer would not be
necessary when using anhydrous ammonia

Selective catalytic reduction of NO x using ammonia as the reducing agent was pat-
ented in the United States by the Engelhard Corporation in 1957. Development of SCR
technology continued in Japan and the US in the early 1960s with research focusing on
less expensive and more durable catalyst agents. The first large-scale SCR was installed
by the IHI Corporation in 1978.

Commercial selective catalytic reduction systems are typically found on large utility boilers,
industrial boilers, and municipal solid waste boilers and have been shown to reduce NO
x by 70-95%. More recent applications include diesel engines, such as those found on
large ships, diesel locomotives, gas turbines, and even automobiles.

Chemistry

The NO_x reduction reaction takes place as the gases pass through the catalyst chamber.
Before entering the catalyst chamber the ammonia, or other reductant (such as urea), is
injected and mixed with the gases. The chemical equation for a stoichiometric reaction
using either anhydrous or aqueous ammonia for a selective catalytic reduction process is:

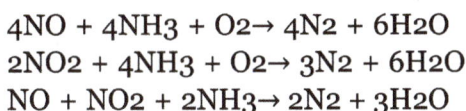

$$4NO + 4NH3 + O2 \rightarrow 4N2 + 6H2O$$
$$2NO2 + 4NH3 + O2 \rightarrow 3N2 + 6H2O$$
$$NO + NO2 + 2NH3 \rightarrow 2N2 + 3H2O$$

With several secondary reactions:

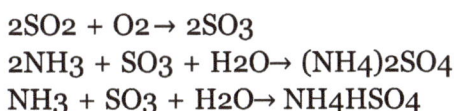

$$2SO2 + O2 \rightarrow 2SO3$$
$$2NH3 + SO3 + H2O \rightarrow (NH4)2SO4$$
$$NH3 + SO3 + H2O \rightarrow NH4HSO4$$

The reaction for urea instead of either anhydrous or aqueous ammonia is:

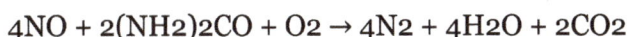

$$4NO + 2(NH2)2CO + O2 \rightarrow 4N2 + 4H2O + 2CO2$$

The ideal reaction has an optimal temperature range between 630 and 720 K, but can operate from 500 to 720 K with longer residence times. The minimum effective temperature depends on the various fuels, gas constituents, and catalyst geometry. Other possible reductants include cyanuric acid and ammonium sulfate.

Catalysts

SCR catalysts are made from various ceramic materials used as a carrier, such as titanium oxide, and active catalytic components are usually either oxides of base metals (such as vanadium, molybdenum and tungsten), zeolites, or various precious metals. Another catalyst based on activated carbon was also developed which is applicable for the removal of NOx at low temperatures. Each catalyst component has advantages and disadvantages.

Base metal catalysts, such as the vanadium and tungsten, lack high thermal durability, but are less expensive and operate very well at the temperature ranges most commonly seen in industrial and utility boiler applications. Thermal durability is particularly important for automotive SCR applications that incorporate the use of a diesel particulate filter with forced regeneration. They also have a high catalysing potential to oxidize SO_2 into SO_3, which can be extremely damaging due to its acidic properties.

Zeolite catalysts have the potential to operate at substantially higher temperature than base metal catalysts; they can withstand prolonged operation at temperatures of 900 K and transient conditions of up to 1120 K. Zeolites also have a lower potential for potentially damaging SO_2 oxidation.

Iron- and copper-exchanged zeolite urea SCRs have been developed with approximately equal performance to that of vanadium-urea SCRs if the fraction of the NO 2 is 20% to 50% of the total NOx. The two most common designs of SCR catalyst geometry used today are honeycomb and plate. The honeycomb form usually is an extruded ceramic applied homogeneously throughout the ceramic carrier or coated on the substrate. Like the various types of catalysts, their configuration also has advantages and disadvantages. Plate-type catalysts have lower pressure drops and are less susceptible to plugging and fouling than the honeycomb types, but plate configurations are much larger and more expensive. Honeycomb configurations are smaller than plate types, but have higher pressure drops and plug much more easily. A third type is corrugated, comprising only about 10% of the market in power plant applications.

Reductants

Several reductants are currently used in SCR applications including anhydrous ammonia, aqueous ammonia or urea. All those three reductants are widely available in large quantities.

Pure anhydrous ammonia is extremely toxic and difficult to safely store, but needs no further conversion to operate within an SCR. It is typically favoured by large industrial SCR operators. Aqueous ammonia must be vaporized in order to be used, but it is substantially safer to store and transport than anhydrous ammonia. Urea is the safest to store, but requires conversion to ammonia through thermal decomposition in order to be used as an effective reductant.

Limitations

SCR systems are sensitive to contamination and plugging resulting from normal operation or abnormal events. Many SCRs are given a finite life due to known amounts of contaminants in the untreated gas. The large majority of catalyst on the market is of porous construction. A clay planting pot is a good example of what SCR catalyst feels like. This porosity is what gives the catalyst the high surface area essential for reduction of NOx. However, the pores are easily plugged by a variety of compounds present in combustion/fuel gas. Some examples of plugging contaminates are: fine particulate, ammonia sulfur compounds, ammonium bisulfate (ABS) and silicon compounds. Many of these contaminants can be removed while the unit is on line, for example by sootblowers. The unit can also be cleaned during a turnaround or by raising the exhaust temperature. Of more concern to SCR performance is poisons, which will destroy the chemistry of the catalyst and render the SCR ineffective at NOx reduction or cause unwanted oxidation of ammonia (forming more NOx). Some of these poisons include: halogens, alkaline metals, arsenic, phosphorus, antimony, chrome, copper.

Most SCRs require tuning to properly perform. Part of tuning involves ensuring a proper distribution of ammonia in the gas stream and uniform gas velocity through the catalyst. Without tuning, SCRs can exhibit inefficient NOx reduction along with excessive ammonia slip due to not utilizing the catalyst surface area effectively. Another facet of tuning involves determining the proper ammonia flow for all process conditions. Ammonia flow is in general controlled based on NOx measurements taken from the gas stream or preexisting performance curves from an engine manufacturer (in the case of gas turbines and reciprocating engines). Typically, all future operating conditions must be known beforehand to properly design and tune an SCR system.

Ammonia slip is an industry term for ammonia passing through the SCR un-reacted. This occurs when ammonia is over-injected into gas stream, temperatures are too low for ammonia to react, or catalyst has degraded.

Temperature is one of the largest limitations of SCR. Gas turbines, cars, and diesel engines all have a period during a start-up where exhaust temperatures are too cool for NOx reduction to occur.

Power Plants

In power stations, the same basic technology is employed for removal of NO x from the flue gas of boilers used in power generation and industry. In general, the SCR unit is located between the furnace economizer and the air heater, and the ammonia is injected into the catalyst chamber through an ammonia injection grid. As in other SCR applications, the temperature of operation is critical. Ammonia slip is also an issue with SCR technology used in power plants.

Other issues that must be considered in using SCR for NOx control in power plants are the formation of ammonium sulfate and ammonium bisulfate due to the sulfur content of the fuel as well as the undesirable catalyst-caused formation of SO_3 from the SO_2 and O_2 in the flue gas.

A further operational difficulty in coal-fired boilers is the binding of the catalyst by fly ash from the fuel combustion. This requires the usage of sootblowers, sonic horns, and careful design of the ductwork and catalyst materials to avoid plugging by the fly ash. SCR catalysts have a typical operational lifetime of about 16,000-40,000 hours in coal-fired power plants, depending on the flue gas composition, and up to 80,000 hours in cleaner gas-fired power plants.

SCR and EPA 2010

Hino truck and its Standardized SCR Unit which combines SCR with Diesel Particulate Active Reduction (DPR). DPR is a diesel particulate filtration system with regeneration process that uses late fuel injection to control exhaust temperature to burn off soot.

Diesel engines manufactured on or after January 1, 2010 are required to meet lowered NOx standards for the US market.

All of the heavy-duty engine (Class 7-8 trucks) manufacturers except for Navistar International and Caterpillar Inc. continuing to manufacture engines after this date have chosen to use SCR. This includes Detroit Diesel (DD13, DD15, and DD16 models), Cummins (ISX, ISL9, ISB6.7, and ISC8.3 line), PACCAR, and Volvo/Mack. These engines require the periodic addition of diesel exhaust fluid (DEF, a urea solution) to enable the process. DEF is available in a bottle from most truck stops, and some provide bulk DEF dispensers near diesel fuel pumps. Caterpillar and Navistar had initially chosen to use enhanced exhaust gas recirculation (EEGR) to comply with the Environmental Protection Agency (EPA) standards, but in July 2012 Navistar announced it would be pursuing SCR technology for its engines, except on the MaxxForce 15 which was to be discontinued.

Daimler AG and Volkswagen have used SCR technology in some of their passenger diesel cars.

Exhaust Gas Recirculation

EGR valve the top of box on top of the inlet manifold of a Saab H engine in a 1987 Saab 90

In internal combustion engines, exhaust gas recirculation (EGR) is a nitrogen oxide (NO_x) emissions reduction technique used in petrol/gasoline and diesel engines. EGR works by recirculating a portion of an engine's exhaust gas back to the engine cylinders. This dilutes the O_2 in the incoming air stream and provides gases inert to combustion to act as absorbents of combustion heat to reduce peak in-cylinder temperatures. NOx is produced in a narrow band of high cylinder temperatures and pressures.

In a gasoline engine, this inert exhaust displaces the amount of combustible matter in the cylinder. In a diesel engine, the exhaust gas replaces some of the excess oxygen in the pre-combustion mixture. Because NOx forms primarily when a mixture of nitrogen and oxygen is subjected to high temperature, the lower combustion chamber temperatures caused by EGR reduces the amount of NOx the combustion generates (though at some loss of engine efficiency). Gasses re-introduced from EGR systems will also contain near equilibrium concentrations of NOx and CO; the small fraction initially within the combustion chamber inhibits the total net production of these and other pollutants when sampled on a time average. Most modern engines now require exhaust gas recirculation to meet emissions standards.

History

The first EGR systems were crude; some were as simple as an orifice jet between the exhaust and intake tracts which admitted exhaust to the intake tract whenever the engine was running. Difficult starting, rough idling, and reduced performance and fuel economy resulted. By 1973, an EGR valve controlled by manifold vacuum opened or closed to admit exhaust to the intake tract only under certain conditions. Control systems grew more sophisticated as automakers gained experience; Chrysler's "Coolant Controlled Exhaust Gas Recirculation" system of 1973 exemplified this evolution: a coolant temperature sensor blocked vacuum to the EGR valve until the engine reached normal operating temperature. This prevented driveability problems due to unnecessary exhaust induction; NOx forms under elevated temperature conditions generally not present with a cold engine. Moreover, the EGR valve was controlled, in part, by vacuum drawn from the carburetor's venturi, which allowed more precise constraint of EGR flow to only those engine load conditions under which NOx is likely to form. Later, backpressure transducers were added to the EGR valve control to further tailor EGR flow to engine load conditions. Most modern engines now need exhaust gas recirculation to meet emissions standards. However, recent innovations have led to the development of engines that do not require them. The 3.6 Chrysler Pentastar engine is one example that does not require EGR.

EGR in Spark-ignited Engines

The exhaust gas, added to the fuel, oxygen, and combustion products, increases the specific heat capacity of the cylinder contents, which lowers the adiabatic flame temperature.

In a typical automotive spark-ignited (SI) engine, 5% to 15% of the exhaust gas is routed back to the intake as EGR. The maximum quantity is limited by the need of the mixture to sustain a continuous flame front during the combustion event; excessive EGR in poorly set up applications can cause misfires and partial burns. Although EGR does measurably slow combustion, this can largely be compensated for by advancing spark timing. The impact of EGR on engine efficiency largely depends on the specific engine design, and sometimes leads to a compromise between efficiency and NOx emissions. A properly operating EGR can theoretically increase the efficiency of gasoline engines via several mechanisms:

- Reduced throttling losses. The addition of inert exhaust gas into the intake system means that for a given power output, the throttle plate must be opened further, resulting in increased inlet manifold pressure and reduced throttling losses.

- Reduced heat rejection. Lowered peak combustion temperatures not only reduces NOx formation, it also reduces the loss of thermal energy to combustion chamber surfaces, leaving more available for conversion to mechanical work during the expansion stroke.

- Reduced chemical dissociation. The lower peak temperatures result in more of the released energy remaining as sensible energy near TDC (Top Dead-Center), rather than being bound up (early in the expansion stroke) in the dissociation of combustion products. This effect is minor compared to the first two.

EGR is typically not employed at high loads because it would reduce peak power output. This is because it reduces the intake charge density. EGR is also omitted at idle (low-speed, zero load) because it would cause unstable combustion, resulting in rough idle.

Since the EGR system recirculates a portion of exhaust gases, over time the valve can become clogged with carbon deposits that prevent it from operating properly. Clogged EGR valves can sometimes be cleaned, but replacement is necessary if the valve is faulty.

In Diesel Engines

In modern diesel engines, the EGR gas is cooled with a heat exchanger to allow the introduction of a greater mass of recirculated gas. Unlike SI engines, diesels are not limited by the need for a contiguous flamefront; furthermore, since diesels always operate with excess air, they benefit from EGR rates as high as 50% (at idle, when there is otherwise a large excess of air) in controlling NOx emissions. Exhaust recirculated back into the cylinder can increase engine wear as carbon particulate wash past the rings and into the oil.

Since diesel engines are unthrottled, EGR does not lower throttling losses in the way that it does for SI engines. Exhaust gas—largely nitrogen, carbon dioxide, and water vapor—has a higher specific heat than air, so it still serves to lower peak combustion temperatures. However, adding EGR to a diesel reduces the specific heat ratio of the combustion gases in the power stroke. This reduces the amount of power that can be extracted by the piston. EGR also tends to reduce the amount of fuel burned in the power stroke. This is evident by the increase in particulate emissions that corresponds to an increase in EGR.

Particulate matter (mainly carbon) that is not burned in the power stroke is wasted energy. Stricter regulations on particulate matter (PM) call for further emission controls to be introduced to compensate for the PM emissions increase caused by EGR. The most common is a diesel particulate filter in the exhaust system which cleans the exhaust but causes a constant minor reduction in fuel efficiency due to the back pressure created. The nitrogen dioxide component of NOx emissions is the primary oxidizer of the soot caught in the DPF at normal operating temperatures. This process is known as passive regeneration. Increasing EGR rates cause passive regeneration to be less effective at managing the PM loading in the DPF. This necessitates periodic active regeneration of the DPF by burning diesel fuel in the oxidation catalyst in order to significantly increase exhaust gas temperatures through the DPF to the point where PM is quickly burned by the residual oxygen in the exhaust.

By feeding the lower oxygen exhaust gas into the intake, diesel EGR systems lower combustion temperature, reducing emissions of NOx. This makes combustion less efficient, compromising economy and power. The normally "dry" intake system of a diesel engine is now subject to fouling from soot, unburned fuel and oil in the EGR bleed, which has little effect on airflow. However, when combined with oil vapor from a PCV system, can cause buildup of sticky tar in the intake manifold and valves. It can also cause problems with components such as swirl flaps, where fitted. Diesel EGR also increases soot production, though this was masked in the US by the simultaneous introduction of diesel particulate filters. EGR systems can also add abrasive contaminants and increase engine oil acidity, which in turn can reduce engine longevity.

Though engine manufacturers have refused to release details of the effect of EGR on fuel economy, the EPA regulations of 2002 that led to the introduction of cooled EGR were associated with a 3% drop in engine efficiency, bucking a trend of a .5% a year increase.

EGR Implementations

Usually, an engine recirculates exhaust gas by piping it from the exhaust manifold to the inlet manifold. This design is called *external* EGR. A control valve (the EGR valve) within the circuit regulates and times the gas flow. Some engines incorporate a camshaft with relatively large overlap during which both the intake valve and the exhaust valve are open, thus trapping exhaust gas within the cylinder by not fully expelling it during the exhaust stroke. A form of internal EGR is used in the rotary Atkinson cycle engine.

EGR can also be implemented by using a variable geometry turbocharger (VGT) which uses variable inlet guide vanes to build sufficient backpressure in the exhaust manifold. For EGR to flow, a pressure difference is required across the intake and exhaust manifold and this is created by the VGT.

Another method that has been experimented with, is using a throttle in a turbocharged diesel engine to decrease the intake pressure, thereby initiating EGR flow.

Modern systems utilizing electronic engine control computers, multiple control inputs, and servo-driven EGR valves typically *improve* performance/efficiency with no impact on drivability and function.

In most modern engines, a faulty or disabled EGR system will cause the computer to display a check engine light and the vehicle to fail an emissions test. The check light can be remedied by ECU remapping.

Biofilter

Biosolids composting plant biofilter mound - note sprinkler
visible front right to maintain proper moisture level for optimum functioning

Biofiltration is a pollution control technique using a bioreactor containing living material to capture and biologically degrade pollutants. Common uses include processing waste water, capturing harmful chemicals or silt from surface runoff, and microbiotic oxidation of contaminants in air.

Examples of biofiltration include;

- Bioswales, biostrips, biobags, bioscrubbers, and trickling filters

- Constructed wetlands and natural wetlands

- Slow sand filters

- Treatment ponds

- Green belts

- Green walls

- Riparian zones, riparian forests, bosques

Control of Air Pollution

When applied to air filtration and purification, biofilters use microorganisms to remove air pollution. The air flows through a packed bed and the pollutant transfers into a thin biofilm on the surface of the packing material. Microorganisms, including bacteria and fungi are immobilized in the biofilm and degrade the pollutant. Trickling filters and bioscrubbers rely on a biofilm and the bacterial action in their recirculating waters.

The technology finds greatest application in treating malodorous compounds and water-soluble volatile organic compounds (VOCs). Industries employing the technology include food and animal products, off-gas from wastewater treatment facilities, pharmaceuticals, wood products manufacturing, paint and coatings application and manufacturing and resin manufacturing and application, etc. Compounds treated are typically mixed VOCs and various sulfur compounds, including hydrogen sulfide. Very large airflows may be treated and although a large area (footprint) has typically been required—a large biofilter (>200,000 acfm) may occupy as much or more land than a football field—this has been one of the principal drawbacks of the technology. Engineered biofilters, designed and built since the early 1990s, have provided significant footprint reductions over the conventional flat-bed, organic media type.

Air cycle system at biosolids composting plant. Large duct in foreground is exhaust air into biofilter shown in next photo

One of the main challenges to optimum biofilter operation is maintaining proper moisture throughout the system. The air is normally humidified before it enters the bed with a watering (spray) system, humidification chamber, bioscrubber, or biotrickling filter. Properly maintained, a natural, organic packing media like peat, vegetable mulch, bark or wood chips may last for several years but engineered, combined natural organic and synthetic component packing materials will generally last much longer, up to 10 years. A number of companies offer these types or proprietary packing materials and multi-year guarantees, not usually provided with a conventional compost or wood chip bed biofilter.

Although widely employed, the scientific community is still unsure of the physical phenomena underpinning biofilter operation, and information about the microorganisms involved continues to be developed. A biofilter/bio-oxidation system is a fairly simple device to construct and operate and offers a cost-effective solution provided the pollutant is biodegradable within a moderate time frame (increasing residence time = increased size and capital costs), at reasonable concentrations (and lb/hr loading rates) and that the airstream is at an organism-viable temperature. For large volumes of air, a biofilter may be the only cost-effective solution. There is no secondary pollution (unlike the case of incineration where additional CO_2 and NO_x are produced from burning fuels) and degradation products form additional biomass, carbon dioxide and water. Media irrigation water, although many systems recycle part of it to reduce operating costs, has a moderately high biochemical oxygen demand (BOD) and may require treatment before disposal. However, this "blowdown water", necessary for proper maintenance of any bio-oxidation system, is generally accepted by municipal publicly owned treatment works without any pretreatment.

Biofilters are being utilized in Columbia Falls, Montana at Plum Creek Timber Company's fiberboard plant. The biofilters decrease the pollution emitted by the manufacturing process and the exhaust emitted is 98% clean. The newest, and largest, biofilter addition to Plum Creek cost $9.5 million, yet even though this new technology is expensive, in the long run it will cost less overtime than the alternative exhaust-cleaning incinerators fueled by natural gas (which are not as environmentally friendly). The biofilters use trillions of microscopic bacteria that cleanse the air being released from the plant.

Water Treatment

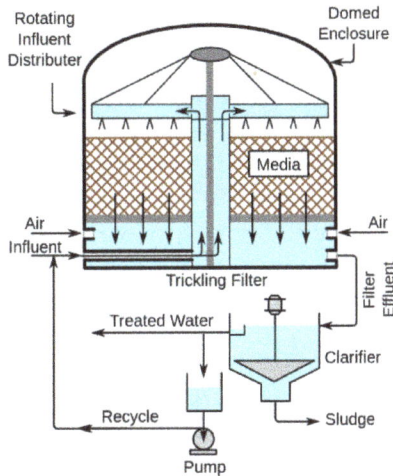

A typical complete trickling filter system for treating wastewaters.

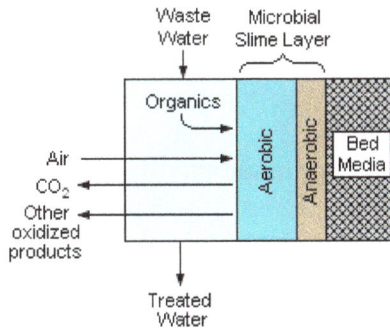

Image 1: A schematic cross-section of the contact face of the bed media in a trickling filter.

Biofiltration was first introduced in England in 1893 as a trickling filter for wastewater treatment and has since been successfully used for the treatment of different types of water. Biological treatment has been used in Europe to filter surface water for drinking purposes since the early 1900s and is now receiving more interest worldwide. Biofiltration is also common in wastewater treatment, aquaculture and greywater recycling, as a way to minimize water replacement while increasing water quality.

Biofiltration Process

Biofilter installation at a commercial composting facility.

A biofilter is a bed of media on which microorganisms attach and grow to form a biological layer called biofilm. Biofiltration is thus usually referred to as a fixed–film process. Generally, the biofilm is formed by a community of different microorganisms (bacteria, fungi, yeast, etc.), macro-organisms (protozoa, worms, insect's larvae, etc.) and extracellular polymeric substances (EPS) (Flemming and Wingender, 2010). The aspect of the biofilm is usually slimy and muddy.

Water to be treated can be applied intermittently or continuously over the media, via upflow or downflow. Typically, a biofilter has two or three phases, depending on the feeding strategy (percolating or submerged biofilter):

- a solid phase (media);

- a liquid phase (water);

- a gaseous phase (air).

Organic matter and other water components diffuse into the biofilm where the treatment occurs, mostly by biodegradation. Biofiltration processes are usually aerobic, which means that microorganisms require oxygen for their metabolism. Oxygen can be supplied to the biofilm, either concurrently or countercurrently with water flow. Aeration occurs passively by the natural flow of air through the process (three phase biofilter) or by forced air supplied by blowers.

Microorganisms' activity is a key-factor of the process performance. The main influencing factors are the water composition, the biofilter hydraulic loading, the type of media, the feeding strategy (percolation or submerged media), the age of the biofilm, temperature, aeration, etc.

Types of Filtering Media

Originally, biofilter was developed using rock or slag as filter media, but different types of material are used today. These materials are categorized as inorganic media (sand, gravel, geotextile, different shapes of plastic media, glass beads, etc.) and organic media (peat, wood chips, coconut shell fragments, compost, etc.)

Advantages

Although biological filters have simple superficial structures, their internal hydrodynamics and the microorganisms' biology and ecology are complex and variable. These characteristics confer robustness to the process. In other words, the process has the capacity to maintain its performance or rapidly return to initial levels following a period of no flow, of intense use, toxic shocks, media backwash (high rate biofiltration processes), etc.

The structure of the biofilm protects microorganisms from difficult environmental conditions and retains the biomass inside the process, even when conditions are not optimal for its growth. Biofiltration processes offer the following advantages: (Rittmann et al., 1988):

- Because microorganisms are retained within the biofilm, biofiltration allows the development of microorganisms with relatively low specific growth rates;

- Biofilters are less subject to variable or intermittent loading and to hydraulic shock;

- Operational costs are usually lower than for activated sludge;

- Final treatment result is less influenced by biomass separation since the biomass concentration at the effluent is much lower than for suspended biomass processes;

- Attached biomass becomes more specialized (higher concentration of relevant organisms) at a given point in the process train because there is no biomass return.

Drawbacks

Because filtration and growth of biomass leads to an accumulation of matter in the filtering media, this type of fixed-film process is subject to clogging and flow channeling. Depending on the type of application and on the media used for microbial growth, clogging can be controlled using physical and/or chemical methods. Whenever possible, backwash steps can be implemented using air and/or water to disrupt the biomat and recover flow. Chemicals such as oxidizing (peroxide, ozone) or biocide agents can also be used.

Drinking Water

For drinking water, biological water treatment involves the use of naturally-occurring microorganisms in the surface water to improve water quality. Under optimum conditions, including relatively low turbidity and high oxygen content, the organisms break down material in the water and thus improve water quality. Slow sand filters or carbon filters are used to provide a support on which these microorganisms grow. These biological treatment systems effectively reduce water-borne diseases, dissolved organic carbon, turbidity and color in surface water, thus improving overall water quality.

Wastewater

Biofiltration is used to treat wastewater from a wide range of sources, with varying organic compositions and concentrations. Many examples of biofiltration applications are described in the literature. As a non-exhaustive list of applications, and notwithstanding the type of media, biofilters were developed and commercialized for the treatment of animal wastes, landfill leachates, dairy wastewater, domestic wastewater.

This process is versatile as it can be adapted to small flows (< 1 m3/d), such as onsite domestic wastewater as well as to flows generated by a municipality (> 240 000 m3/d). For decentralized domestic wastewater production, such as for isolated dwellings, it has been demonstrated that there are important daily, weekly and yearly fluctuations of hydraulic and organic production rates related to modern families' lifestyle. In this context, a biofilter located after a septic tank constitutes a robust process able to sustain the variability observed without compromising the treatment performance.

Use in Aquaculture

The use of biofilters is common in closed aquaculture systems, such as recirculating aquaculture systems (RAS). Many designs are used, with different benefits and drawbacks, however the function is the same: reducing water exchanges by converting ammonia to nitrate. Ammonia (NH_4^+ and NH_3) originates from the brachial excretion from the gills of aquatic animals and from the decomposition of organic matter. As ammonia-N is highly toxic, this is converted to a less toxic form of nitrite (by *Nitrosomonas* sp.) and then to an even less toxic form of nitrate (by *Nitrobacter* sp.). This "nitrification" process requires oxygen (aerobic conditions), without which the biofilter can crash. Furthermore, as this nitrification cycle produces H^+, the pH can decrease which necessitates the use of buffers such as lime.

Fluidized Bed Concentrator

A fluidized bed concentrator (FBC) is an industrial process for the treatment of exhaust air. The system uses a bed of activated carbon beads to adsorb volatile organic compounds (VOCs) from the exhaust gas. Evolving from the previous fixed-bed and carbon rotor concentrators, the FBC system forces the VOC-laden air through several perforated steel trays, increasing the velocity of the air and allowing the sub-millimeter carbon beads to fluidize, or behave as if suspended in a liquid. This increases the surface area of the carbon-gas interaction, making it more effective at capturing VOCs.

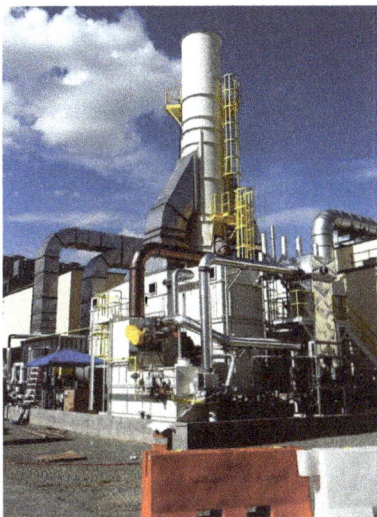

A Fluidized Bed Concentrator for VOC control at Honda Manufacturing of Alabama.

Components

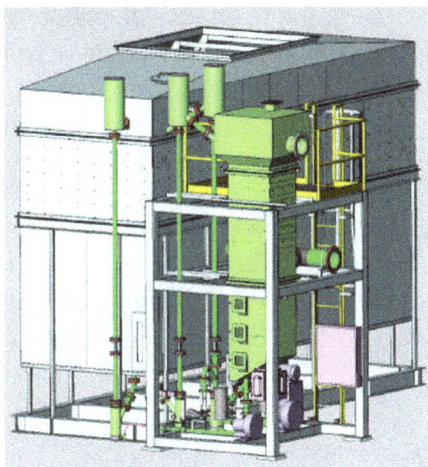

A 3-D design of the fluidized bed concentrator in Solidworks.

The fluidized bed concentrator consists of five primary components:

- Adsorption tower

- Desorption tower

- Thermal oxidizer

- Carbon transport system

- Process fans: Inlet Adsorber, Inlet Desorber, Outlet Oxidizer to Stack

How It Works

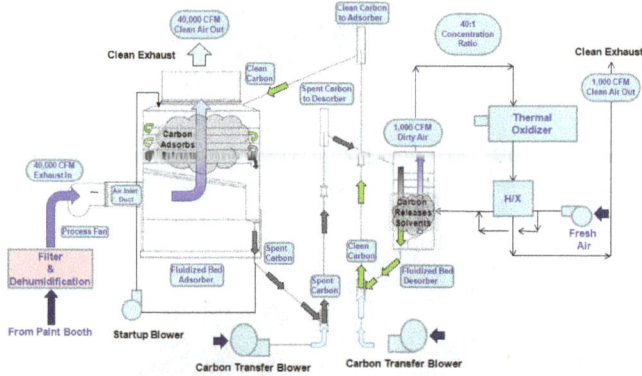

A flow schematic of process gas in the Fluidized Bed Concentrator system.

Industrial Processes requiring ventilation, including paint booths, printing, and chemical production, exhaust the ventilated air to the fluidized bed concentrator (FBC) at room temperature. The air first passes into the Adsorption tower, where it moves through six perforated trays of clean carbon beads. The Bead activated carbon (BAC) fluidizes in the trays and captures the VOCs as they intermix.

The saturated carbon beads are passed from the Adsorber tower to the Desorber tower, where the beads are heated to 350 °F and the VOCs are released. Typically the Adsorber tower is many times larger than the Desorber tower, leading to an air volume reduction and an increase in VOC concentration. The ratio of Adsorber size to Desorber size is called the Concentration Ratio, and ranges from 10:1 to 100:1.

The concentrated VOC gas stream is sent from the Desorb tower to a thermal oxidizer, where the organic compounds are heated to 1400 °F and oxidized, or broken down into Carbon Dioxide (CO_2), Water (H_2O), and by-products. In some cases, small amounts of Carbon Monoxide (CO), Nitrogen Oxide (NOx), and other gases are produced.

Emissions and Energy Usage

The primary advantage of the FBC over traditional rotor concentrators lies in its ability to achieve any concentration ratio up to the lower explosive limit (LEL). This allows Honda Alabama's paint shop to switch from oxidizing 100,000 CFM of VOCs in an RTO, to oxidizing only 1,500 CFM of VOCs in a small thermal oxidizer, at a much higher concentration. Reducing the volume of air to be oxidized from 100,000 CFM to 1,500 CFM (66:1 concentration ratio), allows for a much lower energy usage and consequently, fewer CO_2 and NOx emissions.

Members of the Honda Alabama Environmental Air Quality team are
honored for their efforts to reduce CO_2 and NO_x emissions.

"Despite an increase in Line 2 production, Honda is realizing a reduction in plant VOC emissions of nearly 60 metric tons annually as a result of the installation of the FBC system. Also, the new [FBC] system uses approximately 20% of the energy of an RTO system." - Honda Manufacturing of Alabama

Industries Served

The Adsorber tower and stack of a Fluidized Bed Concentrator.

- Paint Finishing
  - o Automotive
  - o Aerospace
  - o Heavy Machinery
  - o Transportation

- Printing

- Chemical production

- Semiconductor

- Food Processing

Known Suppliers

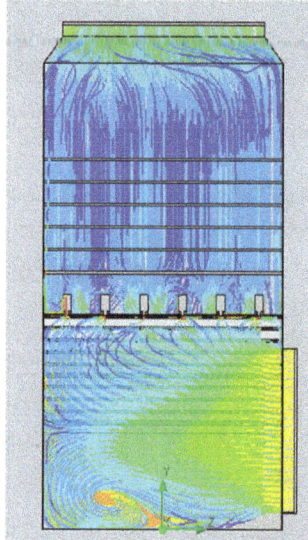

A computational fluid dynamics (CFD) model for airflow inside a FBC Adsorber tower. The air passes through a diffuser and six layers of perforated stainless steel trays.

- Environmental C&C

  o Customers: Sony, Akzo Nobel, Hitachi, Lucent, Panasonic

- TKS Industrial

  o Customers: Toyota, Honda, Ford

- PEI

  o Customers: Nail Polish Manufacturer

Thermal Oxidizer

A thermal oxidizer (also known as thermal oxidiser, or thermal incinerator) is a process unit for air pollution control in many chemical plants that decomposes hazardous gases at a high temperature and releases them into the atmosphere.

Thermal oxidizer installed at a factory.

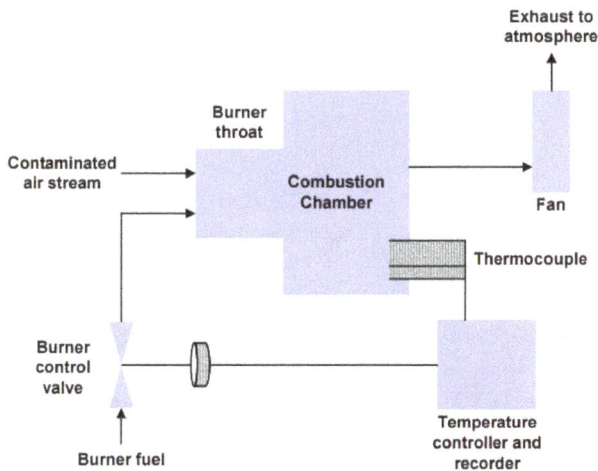

Schematic of a basic thermal oxidizer

Principle

Thermal oxidizers are typically used to destroy hazardous air pollutants (HAPs) and volatile organic compounds (VOCs) from industrial air streams. These pollutants are generally hydrocarbon based and when destroyed via thermal combustion they are chemically oxidized to form CO_2 and H_2O. Three main factors in designing the effective thermal oxidizers are temperature, residence time, and turbulence. The temperature needs to be high enough to ignite the waste gas. Most organic compounds ignite at the temperature between 590 °C (1,094 °F) and 650 °C (1,202 °F). To ensure near destruction of hazardous gases, most basic oxidizers are operated at much higher temperature levels. When catalyst is used, the operating temperature range may be lower. Residence time is to ensure that there is enough time for the combustion reaction to occur. The turbulence factor is the mixture of combustion air with the hazardous gases.

Technologies

Direct Fired Thermal Oxidizer - afterburner

Direct-fired thermal oxidizer using landfill gas as fuel

Regenerative thermal oxidizer (RTO) that is 17000 standard cubic feet per minute, or SCFM for short.

Control center with a programmable logic controller for a RTO.

The simplest technology of thermal oxidation is direct-fired thermal oxidizer. A process stream with hazardous gases is introduced into a firing box through or near the burner and enough residence time is provided to get the desired destruction removal efficiency (DRE) of the VOCs. Most direct-fired thermal oxidizers operate at temperature levels between 980 °C (1,800 °F) and 1,200 °C (2,190 °F) with air flow rates of 0.24 to 24 standard cubic meters per second.

Also called afterburners in the cases where the input gases come from a process where combustion is incomplete, these systems are the least capital intensive, but when applied incorrectly, the operating costs can be devastating because there is no form of heat recovery. These are best applied where there is a very high concentration of VOCs to act as the fuel source (instead of natural gas or oil) for complete combustion at the targeted operating temperature.

Regenerative Thermal Oxidizer (RTO)

One of today's most widely accepted air pollution control technologies across industry is a regenerative thermal oxidizer, commonly referred to as a RTO. RTOs use a ceramic bed which is heated from a previous oxidation cycle to preheat the input gases to partially oxidize them. The preheated gases enter a combustion chamber that is heated by an external fuel source to reach the target oxidation temperature which is in the range between 760 °C (1,400 °F) and 820 °C (1,510 °F). The final temperature may be as high as 1,100 °C (2,010 °F) for applications that require maximum destruction. The air flow rates are 2.4 to 240 standard cubic meters per second.

RTOs are very versatile and extremely efficient – thermal efficiency can reach 95%. They are regularly used for abating solvent fumes, odours, etc. from a wide range of industries. Regenerative Thermal Oxidizers are ideal in a range of low to high VOC concentrations up to 10 g/m³ solvent. There are currently many types Regenerative Thermal Oxidizer on the market with the capabitlity of 99.5+% Volatile Organic Compound (VOC) oxidisation or destruction efficiency. The ceramic heat exchanger(s) in the towers can be designed for thermal efficiencies as high as 97+%.

Ventilation Air Methane Thermal Oxidizer (VAMTOX)

Ventilation air methane thermal oxidizers are used to destroy methane in the exhaust air of underground coal mine shafts. Methane is a greenhouse gas and, when oxidized via thermal combustion, is chemically altered to form CO_2 and H_2O. CO_2 is 25 times less potent than methane when emitted into the atmosphere with regards to global warming. Concentrations of methane in mine ventilation exhaust air of coal and trona mines are very dilute; typically below 1% and often below 0.5%. VAMTOX units have a system of valves and dampers that direct the air flow across one or more ceramic filled bed(s). On start-up, the system preheats by raising the temperature of the heat exchanging ceramic material in the bed(s) at or above the auto-oxidation temperature of methane 1,000 °C (1,830 °F), at which time the preheating system is turned off and mine exhaust air is introduced. Then the methane-filled air reaches the preheated bed(s), releasing the heat from combustion. This heat is then transferred back to the bed(s), thereby maintaining the temperature at or above what is necessary to support auto-thermal operation.

Thermal Recuperative Oxidizer

A less commonly used thermal oxidizer technology is a thermal recuperative oxidizer. Thermal recuperative oxidizers have a primary and/or secondary heat exchanger within the system. A primary heat exchanger preheats the incoming dirty air by recuperating heat from the exiting clean air. This is done by a shell and tube heat exchanger or a plate heat exchanger. As the incoming air passes on one side of the metal tube or plate, hot clean air from the combustion chamber passes on the other side of the tube or plate and heat is transferred to the incoming air through the process of conduction using the metal as the medium of heat transfer. In a secondary heat exchanger the same concept applies for heat transfer, but the air being heated by the outgoing clean process stream is being returned to another part of the plant – perhaps back to the process.

Biomass Fired Thermal Oxidizer

Biomass, such as wood chips, can be used as the fuel for a thermal oxidizer. The biomass is then gasified and the stream with hazardous gases is mixed with the biomassgas in a firing box. Sufficient turbulence, retention time, oxygen content and temperature will ensure destruction of the VOC's. Such biomass fired thermal oxidizer has been installed at Warwick Mills, New Hampshire. The inlet concentrations are between 3000-10.000 ppm VOC. The outlet concentration of VOC are below 3 ppm, thus having a VOC destruction efficiency of 99.8%-99.9%.

Catalytic Oxidizer

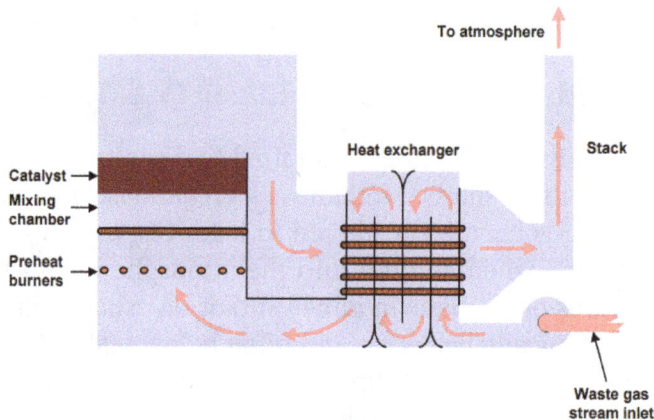

Schematic of Recuperative Catalytic Oxidizer

Catalytic oxidizer (also known as catalytic incinerator) is another category of oxidation systems that is similar to typical thermal oxidizers, but the catalytic oxidizers use a catalyst to promote the oxidation. Catalytic oxidation occurs through a chemical reaction between the VOC hydrocarbon molecules and a precious-metal catalyst bed that is internal to the oxidizer system. A catalyst is a substance that is used to accelerate the rate of a chemical reaction, allowing the reaction to occur in a normal temperature range between 340 °C (644 °F) and 540 °C (1,004 °F).

Regenerative Catalytic Oxidizer (RCO)

The catalyst can be used in a Regenerative Thermal Oxidizer (RTO) to allow lower operating temperatures. This is also called Regenerative Catalytic Oxidizer or RCO. For example, the thermal ignition temperature of carbon monoxide is normally 609 °C (1,128 °F). By utilizing a suitable oxidation catalyst, the ignition temperature can be reduced to around 200 °C (392 °F). This can result in lower operating costs than a RTO. Most systems operate within the 260 °C (500 °F) to 1,000 °C (1,830 °F) degree range. Some systems are designed to operate both as RCOs and RTOs. When these systems are used special design considerations are utilized to reduce the probability of overheating (dilution of inlet gas or recycling), as these high temperatures would deactivate the catalyst, e.g. by sintering of the active material.

Recuperative Catalytic Oxidizer

Catalytic oxidizers can also be in the form of recuperative heat recovery to reduce the fuel requirement. In this form of heat recovery, the hot exhaust gases from the oxidizer pass through an heat exchanger to heat the new incoming air to the oxidizer.

Vapor Recovery

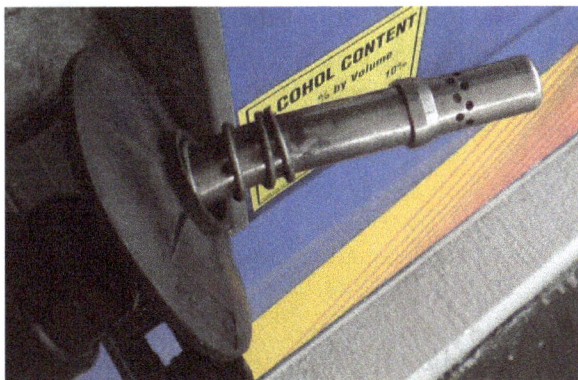

Gas nozzle with vapor recovery

Vapor (or vapour) recovery is the process of recovering the vapors of gasoline or other fuels, so that they do not escape into the atmosphere. This is often done (or required by law) at filling stations, in order to reduce noxious and potentially explosive fumes and pollution.

The negative pressure created in the (underground) tank by the withdrawal is usually used to pull in the vapors. They are drawn-in through holes in the side of the nozzle and travel through special hoses which have a return path.

State Requirements

•   Illinois -

Other Industries

Vapor recovery is also used in the chemical process industry to remove and recover vapors from storage tanks. The vapors are usually either environmentally hazardous, or valuable to be recovered. The process consists of a closed venting system from the storage tank ullage space to a vapor recovery unit (VRU) which will recover the vapors for return to the process or destroy them, usually by oxidation.

Vapor recovery units are also becoming commonly used in the oil and gas industry as a means of recovering natural gas vapor and making it a usable and profitable product. Specifically a newer form of vapor recovery technology, Ejector Vapor Recovery Units create a closed loop system which not only recovers valuable vapor, but also reduces methane and VOC emissions.

In the Australian region vapor recovery has become mandatory in major urban areas. There are 2 categories - VR1 and VR2. VR1 is to be installed in fuel stations that pump less than 500,000 litres annually, VR2 is for fuel quantities over 500,000 litres per annum, or as designated by various EPA bodies.

Vapor recovery towers are also used in the oil and gas industry to provide flash gas recovery at near atmospheric pressure without the potential of oxygen ingress from the top of the storage tanks. The ability to create the vapor flash inside the vapor recovery tower often reduces storage tank emissions to less than 6 tons per year, exempting the tank battery from Quad O reporting requirements.

References

- Rosen (Ed.), Erwin M. (1975). The Peterson automotive troubleshooting & repair manual. Grosset & Dunlap, Inc. ISBN 978-0-448-11946-5.

- Bennett, Sean (2004). Medium/Heavy Duty Truck Engines, Fuel & Computerized Management Systems 2nd Edition, ISBN 1401814999.

- Joseph S. Devinny, Marc A. Deshusses and Todd S. Webster (1999). Biofiltration for Air Pollution Control. Lewis Publishers. ISBN 1-56670-289-5.

- "Thermal Oxidizer". U.S. EPA Technology Transfer Network Clearinghouse for Inventories & Emissions Factors. U.S. Environmental Protection Agency. Retrieved 4 April 2015.

- "Catalytic Oxidizer". U.S. EPA Technology Transfer Network Clearinghouse for Inventories & Emissions Factors. U.S. Environmental Protection Agency. Retrieved 4 April 2015.

- "Honda Manufacturing Alabama Honored As Air Conservationist of the Year". Retrieved 7 November 2014.

- "2011 Dodge Challenger Officially Revealed With 305-HP Pentastar V6". autoguide.com. Retrieved 26 September 2011.

Various Devices to Control Air Pollution

An electrostatic precipitator is a device that is used to filter air; it helps in removing particles of smoke and dust. The alternative devices that are used in controlling air pollution are dust collectors, scrubbers, catalytic converters and air purge systems. The topics discussed in the chapter are of great importance to broaden the existing knowledge on air pollution.

Electrostatic precipitator

An electrostatic precipitator (ESP) is a filtration device that removes fine particles, like dust and smoke, from a flowing gas using the force of an induced electrostatic charge minimally impeding the flow of gases through the unit.

Electrodes inside electrostatic precipitator

In contrast to wet scrubbers which apply energy directly to the flowing fluid medium, an ESP applies energy only to the particulate matter being collected and therefore is very efficient in its consumption of energy (in the form of electricity).

Invention of the Electrostatic Precipitator

The first use of corona discharge to remove particles from an aerosol was by Hohlfeld in 1824. However, it was not commercialized until almost a century later.

In 1907 Frederick Gardner Cottrell, a professor of chemistry at the University of California, Berkeley, applied for a patent on a device for charging particles and then collecting them through electrostatic attraction—the first electrostatic precipitator. Cottrell first applied the device to the collection of sulphuric acid mist and lead oxide fumes emitted from various acid-making and smelting activities. Wine-producing vineyards in northern California were being adversely affected by the lead emissions.

At the time of Cottrell's invention, the theoretical basis for operation was not understood. The operational theory was developed later in Germany, with the work of Walter Deutsch and the formation of the Lurgi company.

Cottrell used proceeds from his invention to fund scientific research through the creation of a foundation called Research Corporation in 1912, to which he assigned the patents. The intent of the organization was to bring inventions made by educators (such as Cottrell) into the commercial world for the benefit of society at large. The operation of Research Corporation is funded by royalties paid by commercial firms after commercialization occurs. Research Corporation has provided vital funding to many scientific projects: Goddard's rocketry experiments, Lawrence's cyclotron, production methods for vitamins A and B_1, among many others.

By a decision of the US Supreme Court, the Corporation had to be split into several entities. The Research Corporation was separated from two commercial firms making the hardware: Research-Cottrell Inc. (operating east of the Mississippi River) and Western Precipitation (operating in the western states). The Research Corporation continues to be active to this day, and the two companies formed to commercialize the invention for industrial and utility applications are still in business as well.

Electrophoresis is the term used for migration of gas-suspended charged particles in a direct-current electrostatic field. Traditional CRT television sets tend to accumulate dust on the screen because of this phenomenon (a CRT is a direct-current machine operating at about 35 kilovolts).

Plate Precipitator

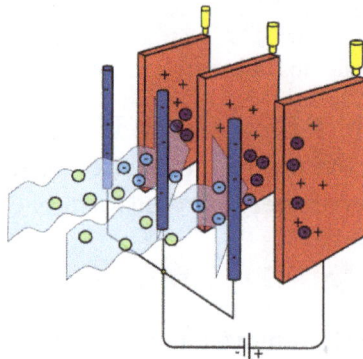

Conceptual diagram of an electrostatic precipitator

The most basic precipitator contains a row of thin vertical wires, and followed by a stack of large flat metal plates oriented vertically, with the plates typically spaced about 1 cm to 18 cm apart, depending on the application. The air stream flows horizontally through the spaces between the wires, and then passes through the stack of plates.

A negative voltage of several thousand volts is applied between wire and plate. If the applied voltage is high enough, an electric corona discharge ionizes the air around the electrodes, which then ionizes the particles in the air stream.

The ionized particles, due to the electrostatic force, are diverted towards the grounded plates. Particles build up on the collection plates and are removed from the air stream.

A two-stage design (separate charging section ahead of collecting section) has the benefit of minimizing ozone production, which would adversely affect health of personnel working in enclosed spaces. For shipboard engine rooms where gearboxes generate an oil mist, two-stage ESP's are used to clean the air, improving the operating environment and preventing buildup of flammable oil fog accumulations. Collected oil is returned to the gear lubricating system.

Collection Efficiency (R)

Precipitator performance is very sensitive to two particulate properties: 1) Electrical resistivity; and 2) Particle size distribution. These properties can be measured economically and accurately in the laboratory, using standard tests. Resistivity can be determined as a function of temperature in accordance with IEEE Standard 548. This test is conducted in an air environment containing a specified moisture concentration. The test is run as a function of ascending or descending temperature, or both. Data is acquired using an average ash layer[further explanation needed] electric field of 4 kV/cm. Since relatively low applied voltage is used and no sulfuric acid vapor is present in the test environment, the values obtained indicate the maximum ash resistivity.

In an ESP, where particle charging and discharging are key functions, resistivity is an important factor that significantly affects collection efficiency. While resistivity is an important phenomenon in the inter-electrode region where most particle charging takes place, it has a particularly important effect on the dust layer at the collection electrode where discharging occurs. Particles that exhibit high resistivity are difficult to charge. But once charged, they do not readily give up their acquired charge on arrival at the collection electrode. On the other hand, particles with low resistivity easily become charged and readily release their charge to the grounded collection plate. Both extremes in resistivity impede the efficient functioning of ESPs. ESPs work best under normal resistivity conditions.

Resistivity, which is a characteristic of particles in an electric field, is a measure of a particle's resistance to transferring charge (both accepting and giving up charges). Resistivity is a function of a particle's chemical composition as well as flue gas operating conditions such as temperature and moisture. Particles can have high, moderate (normal), or low resistivity.

Bulk resistivity is defined using a more general version of Ohm's Law, as given in Equation (1) below:

$$\vec{E} = \rho\vec{j} \tag{1}$$

Where:

```
E is the Electric field strength (V/cm);

j is the Current density (A/cm²); and

ρ is the Resistivity (Ohm-cm)
```

A better way of displaying this would be to solve for resistivity as a function of applied voltage and current, as given in Equation (2) below:

$$\rho = \frac{AV}{Il} \tag{2}$$

Where:

```
ρ = Resistivity (Ohm-cm)

V = The applied DC potential, (Volts);

I = The measured current, (Amperes);

l = The ash layer thickness, (cm); and

A = The current measuring electrode face area, (cm²).
```

Resistivity is the electrical resistance of a dust sample 1.0 cm^2 in cross-sectional area, 1.0 cm thick, and is recorded in units of ohm-cm. A method for measuring resistivity will be described in this article. The table below, gives value ranges for low, normal, and high resistivity.

Resistivity	Range of Measurement
Low	between 10^4 and 10^7 ohm-cm
Normal	between 10^7 and 2×10^{10} ohm-cm
High	above 2×10^{10} ohm-cm

Dust Layer Resistance

resistance affects electrical conditions in the dust layer by a potential electric field (voltage drop) being formed across the layer as negatively charged particles arrive at its surface and leak their electrical charges to the collection plate. At the metal surface of the electrically grounded collection plate, the voltage is zero, whereas at the outer surface of the dust layer, where new particles and ions are arriving, the electrostatic voltage caused by the gas ions can be quite high. The strength of this electric field depends on the resistance and thickness of the dust layer.

In high-resistance dust layers, the dust is not sufficiently conductive, so electrical charges have difficulty moving through the dust layer. Consequently, electrical charges accumulate on and beneath the dust layer surface, creating a strong electric field.

Voltages can be greater than 10,000 volts. Dust particles with high resistance are held too strongly to the plate, making them difficult to remove and causing rapping problems.

In low resistance dust layers, the corona current is readily passed to the grounded collection electrode. Therefore, a relatively weak electric field, of several thousand volts, is maintained across the dust layer. Collected dust particles with low resistance do not adhere strongly enough to the collection plate. They are easily dislodged and become retained in the gas stream.

The electrical conductivity of a bulk layer of particles depends on both surface and volume factors. Volume conduction, or the motions of electrical charges through the interiors of particles, depends mainly on the composition and temperature of the particles. In the higher temperature regions, above 500 °F (260 °C), volume conduction controls the conduction mechanism. Volume conduction also involves ancillary factors, such as compression of the particle layer, particle size and shape, and surface properties.

Volume conduction is represented in the figures as a straight-line at temperatures above 500 °F (260 °C). At temperatures below about 450 °F (230 °C), electrical charges begin to flow across surface moisture and chemical films adsorbed onto the particles. Surface conduction begins to lower the resistivity values and bend the curve downward at temperatures below 500 °F (260 °C).

These films usually differ both physically and chemically from the interiors of the particles owing to adsorption phenomena. Theoretical calculations indicate that moisture films only a few molecules thick are adequate to provide the desired surface conductivity. Surface conduction on particles is closely related to surface-leakage currents occurring on electrical insulators, which have been extensively studied. An interesting practical application of surface-leakage is the determination of dew point by measurement of the current between adjacent electrodes mounted on a glass surface. A sharp rise in current signals the formation of a moisture film on the glass. This method has been used effectively for determining the marked rise in dew point, which occurs when small amounts of sulfuric acid vapor are added to an atmosphere (commercial Dewpoint Meters are available on the market).

The following discussion of normal, high, and low resistance applies to ESPs operated in a dry state; resistance is not a problem in the operation of wet ESPs because of the moisture concentration in the ESP. The relationship between moisture content and resistance is explained later in this work.

Normal Resistivity

As stated above, ESPs work best under normal resistivity conditions. Particles with normal resistivity do not rapidly lose their charge on arrival at the collection electrode. These particles slowly leak their charge to grounded plates and are retained on the collection plates by intermolecular adhesive and cohesive forces. This allows a particulate layer to be built up and then dislodged from the plates by rapping. Within the range of normal dust resistivity (between 10^7 and 2×10^{10} ohm-cm), fly ash is collected more easily than dust having either low or high resistivity.

High Resistivity

If the voltage drop across the dust layer becomes too high, several adverse effects can occur. First, the high voltage drop reduces the voltage difference between the discharge electrode and collection electrode, and thereby reduces the electrostatic field strength used to drive the gas ion-charged particles over to the collected dust layer. As the dust layer builds up, and the electrical charges accumulate on the surface of the dust layer, the voltage difference between the discharge and collection electrodes decreases. The migration velocities of small particles are especially affected by the reduced electric field strength.

Another problem that occurs with high resistivity dust layers is called back corona. This occurs when the potential drop across the dust layer is so great that corona discharges begin to appear in the gas that is trapped within the dust layer. The dust layer breaks down electrically, producing small holes or craters from which back corona discharges occur. Positive gas ions are generated within the dust layer and are accelerated toward the "negatively charged" discharge electrode. The positive ions reduce some of the negative charges on the dust layer and neutralize some of the negative ions on the "charged particles" heading toward the collection electrode. Disruptions of the normal corona process greatly reduce the ESP's collection efficiency, which in severe cases, may fall below 50%. When back corona is present, the dust particles build up on the electrodes forming a layer of insulation. Often this can not be repaired without bringing the unit offline.

The third, and generally most common problem with high resistivity dust is increased electrical sparking. When the sparking rate exceeds the "set spark rate limit," the automatic controllers limit the operating voltage of the field. This causes reduced particle charging and reduced migration velocities toward the collection electrode. High resistivity can generally be reduced by doing the following:

- Adjusting the temperature;

- Increasing moisture content;

- Adding conditioning agents to the gas stream;

- Increasing the collection surface area; and

- Using hot-side precipitators (occasionally and with foreknowledge of sodium depletion).

Thin dust layers and high-resistivity dust especially favor the formation of back corona craters. Severe back corona has been observed with dust layers as thin as 0.1 mm, but a dust layer just over one particle thick can reduce the sparking voltage by 50%. The most marked effects of back corona on the current-voltage characteristics are:

1. Reduction of the spark over voltage by as much as 50% or more;

2. Current jumps or discontinuities caused by the formation of stable back-corona craters; and

3. Large increase in maximum corona current, which just below spark over corona gap may be several times the normal current.

The Figure below and to the left shows the variation in resistivity with changing gas temperature for six different industrial dusts along with three coal-fired fly ashes. The Figure on the right illustrates resistivity values measured for various chemical compounds that were prepared in the laboratory.

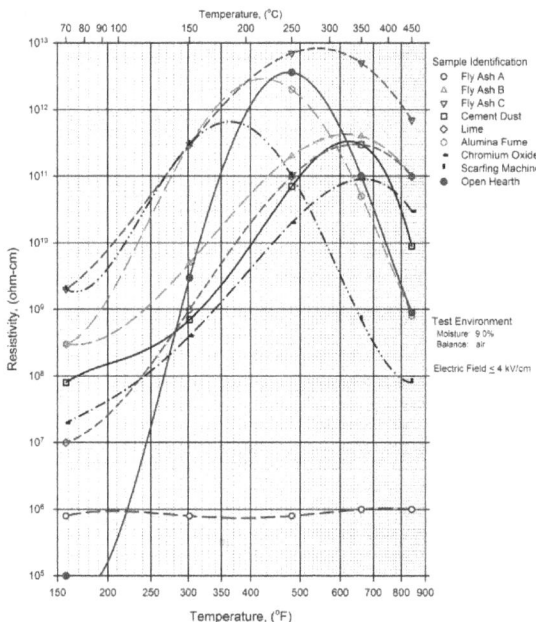

Resistivity Values of Representative Dusts and Fumes From Industrial Plants

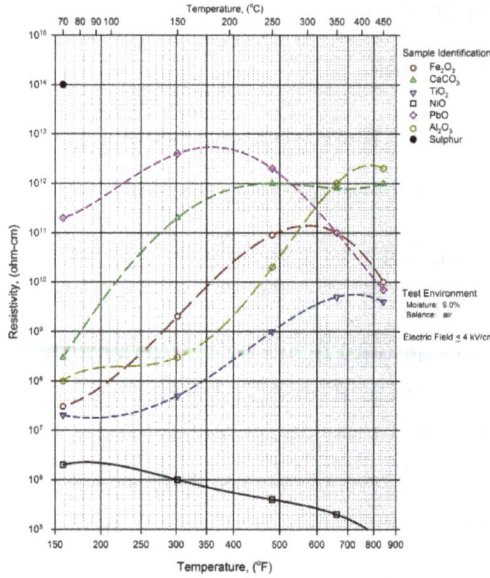

Resistivity Values of Various Chemicals and Reagents as a Function of Temperature

Results for Fly Ash A (in the figure to the left) were acquired in the ascending temperature mode. These data are typical for a moderate to high combustibles content ash. Data for Fly Ash B are from the same sample, acquired during the descending temperature mode.

The differences between the ascending and descending temperature modes are due to the presence of unburned combustibles in the sample. Between the two test modes, the samples are equilibrated in dry air for 14 hours (overnight) at 850 °F (450 °C). This overnight annealing process typically removes between 60% and 90% of any unburned combustibles present in the samples. Exactly how carbon works as a charge carrier is not fully understood, but it is known to significantly reduce the resistivity of a dust.

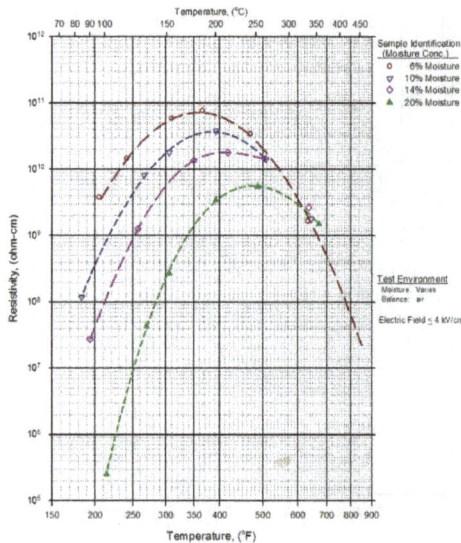

Resistivity Measured as a Function of Temperature in Varying Moisture Concentrations (Humidity)

Carbon can act, at first, like a high resistivity dust in the precipitator. Higher voltages can be required in order for corona generation to begin. These higher voltages can be problematic for the TR-Set controls. The problem lies in onset of corona causing large amounts of current to surge through the (low resistivity) dust layer. The controls sense this surge as a spark. As precipitators are operated in spark-limiting mode, power is terminated and the corona generation cycle re-initiates. Thus, lower power (current) readings are noted with relatively high voltage readings.

The same thing is believed to occur in laboratory measurements. Parallel plate geometry is used in laboratory measurements without corona generation. A stainless steel cup holds the sample. Another stainless steel electrode weight sits on top of the sample (direct contact with the dust layer). As voltage is increased from small amounts (e.g. 20 V), no current is measured. Then, a threshold voltage level is reached. At this level, current surges through the sample... so much so that the voltage supply unit can trip off. After removal of the unburned combustibles during the above-mentioned annealing procedure, the descending temperature mode curve shows the typical inverted "V" shape one might expect.

Low Resistivity

Particles that have low resistivity are difficult to collect because they are easily charged (very conductive) and rapidly lose their charge on arrival at the collection electrode. The particles take on the charge of the collection electrode, bounce off the plates, and become re-entrained in the gas stream. Thus, attractive and repulsive electrical forces that are normally at work at normal and higher resistivities are lacking, and the binding forces to the plate are considerably lessened. Examples of low-resistivity dusts are unburned carbon in fly ash and carbon black.

If these conductive particles are coarse, they can be removed upstream of the precipitator by using a device such as a cyclone mechanical collector.

The addition of liquid ammonia (NH_3) into the gas stream as a conditioning agent has found wide use in recent years. It is theorized that ammonia reacts with H_2SO_4 contained in the flue gas to form an ammonium sulfate compound that increases the cohesivity of the dust. This additional cohesivity makes up for the loss of electrical attraction forces.

The table below summarizes the characteristics associated with low, normal and high resistivity dusts.

The moisture content of the flue gas stream also affects particle resistivity. Increasing the moisture content of the gas stream by spraying water or injecting steam into the duct work preceding the ESP lowers the resistivity. In both temperature adjustment and moisture conditioning, one must maintain gas conditions above the dew point to prevent corrosion problems in the ESP or downstream equipment. The figure to the right shows the effect of temperature and moisture on the resistivity of a cement dust. As the percentage of moisture in the gas stream increases from 6 to 20%, the resistivity of the dust dramatically decreases. Also, raising or lowering the temperature can decrease cement dust resistivity for all the moisture percentages represented.

The presence of SO_3 in the gas stream has been shown to favor the electrostatic precipitation process when problems with high resistivity occur. Most of the sulfur content in the coal burned for combustion sources converts to SO_2. However, approximately 1% of the sulfur converts to SO_3. The amount of SO_3 in the flue gas normally increases with increasing sulfur content of the coal. The resistivity of the particles decreases as the sulfur content of the coal increases.

Resistivity	Range of Measurement	Precipitator Characteristics
Low	between 10^4 and 10^7 ohm-cm	1. Normal operating voltage and current levels unless dust layer is thick enough to reduce plate clearances and cause higher current levels. 2. Reduced electrical force component retaining collected dust, vulnerable to high reentrainment losses. 3. Negligible voltage drop across dust layer. 4. Reduced collection performance due to (2)
Normal	between 10^7 and 2×10^{10} ohm-cm	1. Normal operating voltage and current levels. 2. Negligible voltage drop across dust layer. 3. Sufficient electrical force component retaining collected dust. 4. High collection performance due to (1), (2) and (3)
Marginal to High	between 2×10^{10} and 10^{12} ohm-cm	1. Reduced operating voltage and current levels with high spark rates. 2. Significant voltage loss across dust layer. 3. Moderate electrical force component retaining collected dust. 4. Reduced collection performance due to (1) and (2)
High	above 10^{12} ohm-cm	1. Reduced operating voltage levels; high operating current levels if power supply controller is not operating properly. 2. Very significant voltage loss across dust layer. 3. High electrical force component retaining collected dust. 4. Seriously reduced collection performance due to (1), (2) and probably back corona.

Other conditioning agents, such as sulfuric acid, ammonia, sodium chloride, and soda ash (sometimes as raw trona), have also been used to reduce particle resistivity. Therefore, the chemical composition of the flue gas stream is important with regard to the resistivity of the particles to be collected in the ESP. The table below lists various conditioning agents and their mechanisms of operation.

Conditioning Agent	Mechanism(s) of Action
Sulfur Trioxide and/or Sulfuric Acid	1. Condensation and adsorption on fly ash surfaces. 2. may also increase cohesiveness of fly ash. 3. Reduces resistivity
Ammonia	Mechanism is not clear, various ones proposed; 1. Modifies resistivity. 2. Increases ash cohesiveness. 3. Enhances space charge effect.
Ammonium Sulfate	Little is known about the mechanism; claims are made for the following: 1. Modifies resistivity (depends upon injection temperature). 2. Increases ash cohesiveness. 3. Enhances space charge effect. 4. Experimental data lacking to substantiate which of these is predominant.
Triethylamine	Particle agglomeration claimed; no supporting data.
Sodium Compounds	1. Natural conditioner if added with coal. 2. Resistivity modifier if injected into gas stream.
Compounds of Transition Metals	Postulated that they catalyze oxidation of SO_2 to SO_3; no definitive tests with fly ash to verify this postulation.
Potassium Sulfate and Sodium Chloride	In cement and lime kiln ESPs: 1. Resistivity modifiers in the gas stream. 2. NaCl - natural conditioner when mixed with coal.

If injection of ammonium sulfate occurs at a temperature greater than about 600 °F (320 °C), dissociation into ammonia and sulfur trioxide results. Depending on the ash, SO_2 may preferentially interact with fly ash as SO_3 conditioning. The remainder recombines with ammonia to add to the space charge as well as increase cohesiveness of the ash.

More recently, it has been recognized that a major reason for loss of efficiency of the electrostatic precipitator is due to particle buildup on the charging wires in addition to the collection plates (Davidson and McKinney, 1998). This is easily remedied by making sure that the wires themselves are cleaned at the same time that the collecting plates are cleaned.

Sulfuric acid vapor (SO_3) enhances the effects of water vapor on surface conduction. It is physically adsorbed within the layer of moisture on the particle surfaces. The effects of relatively small amounts of acid vapor can be seen in the figure below and to the right.

The inherent resistivity of the sample at 300 °F (150 °C) is 5×10^{12} ohm-cm. An equilibrium concentration of just 1.9 ppm sulfuric acid vapor lowers that value to about 7×10^9 ohm-cm.

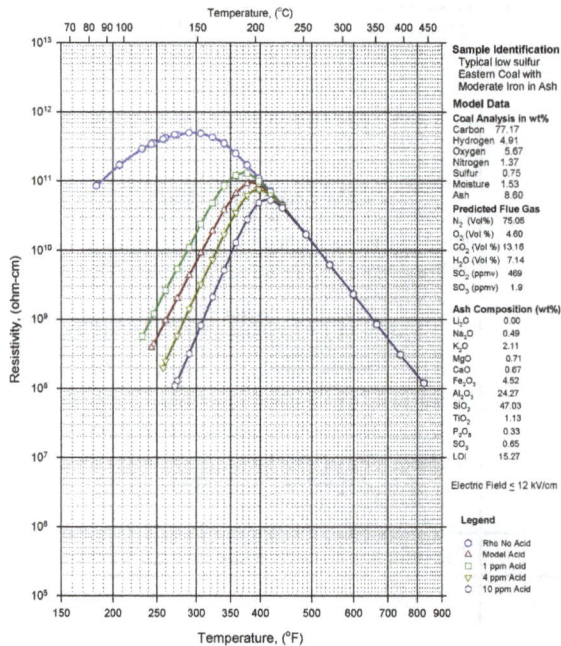

Resistivity Modeled As A Function of Environmental Conditions - Especially Sulfuric Acid Vapor

Modern Industrial Electrostatic Precipitators

ESPs continue to be excellent devices for control of many industrial particulate emissions, including smoke from electricity-generating utilities (coal and oil fired), salt cake collection from black liquor boilers in pulp mills, and catalyst collection from fluidized bed catalytic cracker units in oil refineries to name a few. These devices treat gas volumes from several hundred thousand ACFM to 2.5 million ACFM (1,180 m³/s) in the largest coal-fired boiler applications. For a coal-fired boiler the collection is usually performed downstream of the air preheater at about 160 °C (320 °F) which provides optimal resistivity of the coal-ash particles. For some difficult applications with low-sulfur fuel hot-end units have been built operating above 370 °C (698 °F).

A smokestack at coal-fired Hazelwood Power Station in Victoria,
Australia emits brown smoke when its ESP is shut down

The original parallel plate–weighted wire design has evolved as more efficient (and robust) discharge electrode designs were developed, today focusing on rigid (pipe-frame) discharge electrodes to which many sharpened spikes are attached (barbed wire), maximizing corona production. Transformer-rectifier systems apply voltages of 50–100 kV at relatively high current densities. Modern controls, such as an automatic voltage control, minimize electric sparking and prevent arcing (sparks are quenched within 1/2 cycle of the TR set), avoiding damage to the components. Automatic plate-rapping systems and hopper-evacuation systems remove the collected particulate matter while on line, theoretically allowing ESPs to stay in continuous operation for years at a time.

Electrostatic Air Sampling

Electrostatic precipitators can be used to sample airborne particles or aerosol for analysis purposes. Precipitator designs with a liquid counterelectrode can be used to sample biological particles, e.g. viruses, directly into a small liquid volume to reduce unnecessary sample dilution.

Wet Electrostatic Precipitator

A wet electrostatic precipitator (WESP or wet ESP) operates with water vapor saturated air streams (100% relative humidity). WESPs are commonly used to remove liquid droplets such as sulfuric acid mist from industrial process gas streams. The WESP is also commonly used where the gases are high in moisture content, contain combustible particulate, or have particles that are sticky in nature.

The preferred and most modern type of WESP is a downflow tubular design. This design allows the collected moisture and particulate to form a moving slurry that helps to keep the collection surfaces clean. Plate style and upflow design WESPs are very unreliable and should not be used in applications where particulate is sticky in nature.

Consumer-oriented Electrostatic Air Cleaners

A portable electrostatic air cleaner marketed to consumers

Portable electrostatic air cleaner with cover removed, showing collector plates

Plate precipitators are commonly marketed to the public as air purifier devices or as a permanent replacement for furnace filters, but all have the undesirable attribute of being somewhat messy to clean. A negative side-effect of electrostatic precipitation devices is the potential production of toxic ozone and NO_x. However, electrostatic precipitators offer benefits over other air purifications technologies, such as HEPA filtration, which require expensive filters and can become "production sinks" for many harmful forms of bacteria.

With electrostatic precipitators, if the collection plates are allowed to accumulate large amounts of particulate matter, the particles can sometimes bond so tightly to the metal plates that vigorous washing and scrubbing may be required to completely clean the collection plates. The close spacing of the plates can make thorough cleaning difficult, and the stack of plates often cannot be easily disassembled for cleaning. One solution, suggested by several manufacturers, is to wash the collector plates in a dishwasher.

Some consumer precipitation filters are sold with special soak-off cleaners, where the entire plate array is removed from the precipitator and soaked in a large container overnight, to help loosen the tightly bonded particulates.

A study by the Canada Mortgage and Housing Corporation testing a variety of forced-air furnace filters found that ESP filters provided the best, and most cost-effective means of cleaning air using a forced-air system.

The first portable electrostatic air filter systems for homes was marketed in 1954 by Raytheon.

Dust Collector

Baghouse Dust Collector for Asphalt Plants

Two rooftop dust collectors in Pristina, Kosovo

A dust collector is a system used to enhance the quality of air released from industrial and commercial processes by collecting dust and other impurities from air or gas. Designed to handle high-volume dust loads, a dust collector system consists of a blower, dust filter, a filter-cleaning system, and a dust receptacle or dust removal system. It is distinguished from air cleaners, which use disposable filters to remove dust.

History

Wilhelm Beth

"Beth"-Filter "KS" (1910)

The father of the dust collector was Wilhelm Beth from Lübeck.

Uses

Dust collectors are used in many processes to either recover valuable granular solid or powder from process streams, or to remove granular solid pollutants from exhaust gases prior to venting to the atmosphere. Dust collection is an online process for collecting any process-generated dust from the source point on a continuous basis. Dust collectors may be of single unit construction, or a collection of devices used to separate particulate matter from the process air. They are often used as an air pollution control device to maintain or improve air quality.

Mist collectors remove particulate matter in the form of fine liquid droplets from the air. They are often used for the collection of metal working fluids, and coolant or oil mists. Mist collectors are often used to improve or maintain the quality of air in the workplace environment.

Fume and smoke collectors are used to remove sub-micrometer-size particulates from the air. They effectively reduce or eliminate particulate matter and gas streams from many industrial processes such as welding, rubber and plastic processing, high speed machining with coolants, tempering, and quenching.

Types of Dust Collectors

Five main types of industrial dust collectors are:

- Inertial separators

- Fabric filters

- Wet scrubbers

- Unit collectors

- Electrostatic precipitators

Inertial Separators

Inertial separators separate dust from gas streams using a combination of forces, such as centrifugal, gravitational, and inertial. These forces move the dust to an area where the forces exerted by the gas stream are minimal. The separated dust is moved by gravity into a hopper, where it is temporarily stored.

The three primary types of inertial separators are:

- Settling chambers

- Baffle chambers

- Centrifugal collectors

Neither settling chambers nor baffle chambers are commonly used in the minerals processing industry. However, their principles of operation are often incorporated into the design of more efficient dust collectors.

Settling Chamber

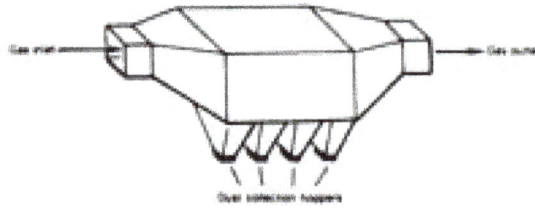

Settling Chamber

A settling chamber consists of a large box installed in the ductwork. The increase of cross section area at the chamber reduces the speed of the dust-filled airstream and heavier particles settle out.

Settling chambers are simple in design and can be manufactured from almost any material. However, they are seldom used as primary dust collectors because of their large space requirements and low efficiency. A practical use is as precleaners for more efficient collectors.

Baffle Chamber

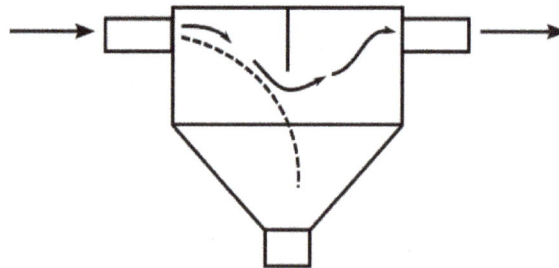

Baffle Chamber
Diagram of a Baffle Chamber

Baffle chambers use a fixed baffle plate that causes the conveying gas stream to make a sudden change of direction. Large-diameter particles do not follow the gas stream but continue into a dead air space and settle. Baffle chambers are used as precleaners

Centrifugal Collectors

Centrifugal collectors use cyclonic action to separate dust particles from the gas stream. In a typical cyclone, the dust gas stream enters at an angle and is spun rapidly. The centrifugal force created by the circular flow throws the dust particles toward the wall of the cyclone. After striking the wall, these particles fall into a hopper located underneath.

The most common types of centrifugal, or inertial, collectors in use today are:-

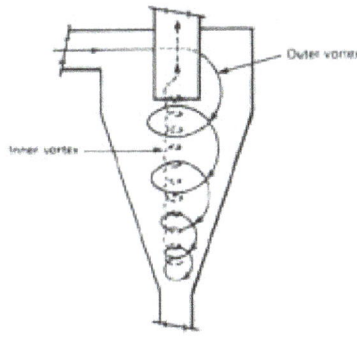

Cyclone

Single-cyclone Separators

They create a dual vortex to separate coarse from fine dust. The main vortex spirals downward and carries most of the coarser dust particles. The inner vortex, created near the bottom of the cyclone, spirals upward and carries finer dust particles.

Multiple-cyclone Separators

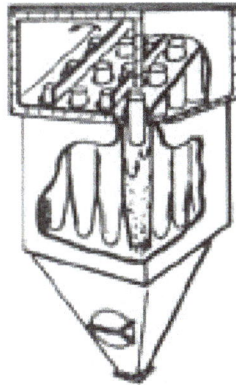

Multiclone

Multiple-cyclone separators consist of a number of small-diameter cyclones, operating in parallel and having a common gas inlet and outlet, as shown in the figure, and operate on the same principle as single cyclone separators—creating an outer downward vortex and an ascending inner vortex.

Multiple-cyclone separators remove more dust than single cyclone separators because the individual cyclones have a greater length and smaller diameter. The longer length provides longer residence time while the smaller diameter creates greater centrifugal force. These two factors result in better separation of dust particulates. The pressure drop of multiple-cyclone separators collectors is higher than that of single-cyclone separators, requiring more energy to clean the same amount of air. A single-chamber cyclone separator of the same volume is more economical, but doesn't remove as much dust.

Cyclone separators are found in all types of power and industrial applications, including pulp and paper plants, cement plants, steel mills, petroleum coke plants, metallurgical plants, saw mills and other kinds of facilities that process dust.

Secondary-air-flow Separators

This type of cyclone uses a secondary air flow, injected into the cyclone to accomplish several things. The secondary air flow increases the speed of the cyclonic action making the separator more efficient; it intercepts the particulate before it reaches the interior walls of the unit; and it forces the separated particulate toward the collection area. The secondary air flow protects the separator from particulate abrasion and allows the separator to be installed horizontally because gravity is not depended upon to move the separated particulate downward.

Fabric Filters

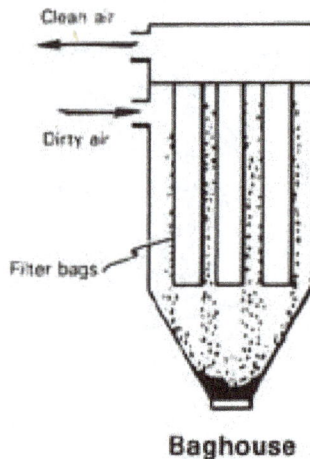

Baghouse

Commonly known as baghouses, fabric collectors use filtration to separate dust particulates from dusty gases. They are one of the most efficient and cost effective types of dust collectors available and can achieve a collection efficiency of more than 99% for very fine particulates.

Dust-laden gases enter the baghouse and pass through fabric bags that act as filters. The bags can be of woven or felted cotton, synthetic, or glass-fiber material in either a tube or envelope shape.

Pre-coating

To ensure the filter bags have a long usage life they are commonly coated with a filter enhancer (pre-coat). The use of chemically inert limestone (calcium carbonate) is most common as it maximises efficiency of dust collection (including fly ash) via formation of what is called a dustcake or coating on the surface of the filter media. This not only traps fine particulates but also provides protection for the bag itself from moisture, and oily or sticky particulates which can bind the filter media. Without a pre-coat the filter bag allows fine particulates to bleed through the bag filter system, especially during start-up, as the bag can only do part of the filtration leaving the finer parts to the filter enhancer dustcake.

Parts

Fabric filters generally have the following parts:

1. Clean plenum

2. Dusty plenum

3. Bag, cage, venturi assembly

4. Tubeplate

5. RAV/SCREW

6. Compressed air header

7. Blow pipe

8. Housing and hopper

Types of Bag Cleaning

Baghouses are characterized by their cleaning method.

Shaking

A rod connecting to the bag is powered by a motor. This provides motion to remove caked-on particles. The speed and motion of the shaking depends on the design of the bag and composition of the particulate matter. Generally shaking is horizontal. The top of the bag is closed and the bottom is open. When shaken, the dust collected on the inside of the bag is freed. During the cleaning process, no dirty gas flows through a bag while the bag is being cleaned. This redirection of air flow illustrates why baghouses must be compartmentalized.

Reverse Air

Air flow gives the bag structure. Dirty air flows through the bag from the inside, allowing dust to collect on the interior surface. During cleaning, gas flow is restricted from a specific compartment. Without the flowing air, the bags relax. The cylindrical bag contains rings that prevent it from completely collapsing under the pressure of the air. A fan blows clean air in the reverse direction. The relaxation and reverse air flow cause the dust cake to crumble and release into the hopper. Upon the completion of the cleaning process, dirty air flow continues and the bag regains its shape.

Pulse Jet

This type of baghouse cleaning (also known as pressure-jet cleaning) is the most common. A high pressure blast of air is used to remove dust from the bag. The blast enters the top of the bag tube, temporarily ceasing the flow of dirty air. The shock of air causes a wave of expansion to travel down the fabric. The flexing of the bag shatters and discharges the dust cake. The air burst is about 0.1 second and it takes about 0.5 seconds for the shock wave to travel down the length of the bag. Due to its rapid release, the blast of air does not interfere with contaminated gas flow. Therefore, pulse-jet baghouses can operate continuously and are not usually compartmentalized. The blast of compressed air must be powerful enough to ensure that the shock wave will travel the entire length of the bag and fracture the dust cake.

Sonic

The least common type of cleaning method is sonic. Shaking is achieved by sonic vibration. A sound generator produces a low frequency sound that causes the bags to vibrate. Sonic cleaning is commonly combined with another method of cleaning to ensure thorough cleaning.

Cartridge Collectors

Cartridge collectors use perforated metal cartridges that contain a pleated, nonwoven filtering media, as opposed to woven or felt bags used in baghouses. The pleated design allows for a greater total filtering surface area than in a conventional bag of the same diameter, The greater filtering area results in a reduced air to media ratio, pressure drop, and overall collector size.

Cartridge collectors are available in single use or continuous duty designs. In single-use collectors, the dirty cartridges are changed and collected dirt is removed while the collector is off. In the continuous duty design, the cartridges are cleaned by the conventional pulse-jet cleaning system.

Wet Scrubbers

Dust collectors that use liquid are known as wet scrubbers. In these systems, the scrubbing liquid (usually water) comes into contact with a gas stream containing dust particles. Greater contact of the gas and liquid streams yields higher dust removal efficiency.

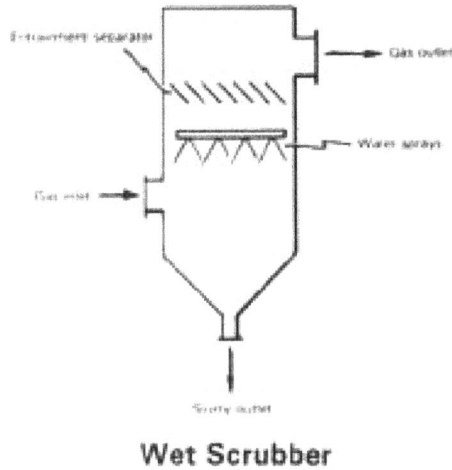

Wet Scrubber

There is a large variety of wet scrubbers; however, all have one of three basic configurations:

1. Gas-humidification - The gas-humidification process agglomerates fine particles, increasing the bulk, making collection easier.

2. Gas-liquid contact - This is one of the most important factors affecting collection efficiency. The particle and droplet come into contact by four primary mechanisms:

    a) Inertial impaction - When water droplets placed in the path of a dust-laden gas stream, the stream separates and flows around them. Due to inertia, the larger dust particles will continue on in a straight path, hit the droplets, and become encapsulated.

    b) Interception - Finer particles moving within a gas stream do not hit droplets directly but brush against and adhere to them.

    c) Diffusion - When liquid droplets are scattered among dust particles, the particles are deposited on the droplet surfaces by Brownian movement, or diffusion. This is the principal mechanism in the collection of submicrometre dust particles.

    d) Condensation nucleation - If a gas passing through a scrubber is cooled below the dewpoint, condensation of moisture occurs on the dust particles. This increase in particle size makes collection easier.

3. Gas-liquid separation - Regardless of the contact mechanism used, as much liquid and dust as possible must be removed. Once contact is made, dust particulates and water droplets combine to form agglomerates. As the agglomerates grow larger, they settle into a collector.

The "cleaned" gases are normally passed through a mist eliminator (demister pads) to remove water droplets from the gas stream. The dirty water from the scrubber system is either cleaned and discharged or recycled to the scrubber. Dust is removed from the scrubber in a clarification unit or a drag chain tank. In both systems solid material settles on the bottom of the tank. A drag chain system removes the sludge and deposits in into a dumpster or stockpile.

Types of Scrubbers

Spray-tower scrubber wet scrubbers may be categorized by pressure drop as follows:

- Low-energy scrubbers (0.5 to 2.5 inches water gauge - 124.4 to 621.9 Pa)

- Low- to medium-energy scrubbers (2.5 to 6 inches water gauge - 0.622 to 1.493 kPa)

- Medium- to high-energy scrubbers (6 to 15 inches water gauge - 1.493 to 3.731 kPa)

- High-energy scrubbers (greater than 15 inches water gauge - greater than 3.731 kPa)

Due to the large number of commercial scrubbers available, it is not possible to describe each individual type here. However, the following sections provide examples of typical scrubbers in each category.

Low-energy Scrubbers

In the simple, gravity-spray-tower scrubber, liquid droplets formed by liquid atomized in spray nozzles fall through rising exhaust gases. Dirty water is drained at the bottom.

These scrubbers operated at pressure drops of 1 to 2 in. water gauge (¼ to ½ kPa) and are approximately 70% efficient on 10 μm particles. Their efficiency is poor below 10 μm. However, they are capable of treating relatively high dust concentrations without becoming plugged.

Low- to Medium-energy Scrubbers

Wet cyclones use centrifugal force to spin the dust particles (similar to a cyclone), and throw the particulates upon the collector's wetted walls. Water introduced from the top to wet the cyclone walls carries these particles away. The wetted walls also prevent dust reentrainment.

Pressure drops for these collectors range from 2 to 8 in. water (½ to 2 kPa), and the collection efficiency is good for 5 μm particles and above.

high-energy Scrubbers co-current-flow Scrubber

Packed-bed scrubbers consist of beds of packing elements, such as coke, broken rock, rings, saddles, or other manufactured elements. The packing breaks down the liquid flow into a high-surface-area film so that the dusty gas streams passing through the bed achieve maximum contact with the liquid film and become deposited on the surfaces of the packing elements. These scrubbers have a good collection efficiency for respirable dust.

Three types of packed-bed scrubbers are-

- Cross-flow scrubbers

- Co-current flow scrubbers

- Counter-current flow scrubbers

Efficiency can be greatly increased by minimizing target size, i.e., using 0.003 in. (0.076 mm) diameter stainless steel wire and increasing gas velocity to more than 1,800 ft/min (9.14 m/s).

High-energy Scrubbers

Venturi scrubbers consist of a venturi-shaped inlet and separator. The dust-laden gases venturi scrubber enter through the venturi and are accelerated to speeds between 12,000 and 36,000 ft/min (60.97-182.83 m/s). These high-gas velocities immediately atomize the coarse water spray, which is injected radially into the venturi throat, into fine droplets. High energy and extreme turbulence promote collision between water droplets and dust particulates in the throat. The agglomeration process between particle and droplet continues in the diverging section of the venturi. The large agglomerates formed in the venturi are then removed by an inertial separator.

Venturi scrubbers achieve very high collection efficiencies for respirable dust. Since efficiency of a venturi scrubber depends on pressure drop, some manufacturers supply a variable-throat venturi to maintain pressure drop with varying gas flows.

Electrostatic Precipitators (ESP)

Electrostatic precipitators use electrostatic forces to separate dust particles from exhaust gases. A number of high-voltage, direct-current discharge electrodes are placed between grounded collecting electrodes. The contaminated gases flow through the passage formed by the discharge and collecting electrodes. Electrostatic precipitators operate on the same principle as home "Ionic" air purifiers.

The airborne particles receive a negative charge as they pass through the ionized field between the electrodes. These charged particles are then attracted to a grounded or positively charged electrode and adhere to it.

The collected material on the electrodes is removed by rapping or vibrating the collecting electrodes either continuously or at a predetermined interval. Cleaning a precipitator can usually be done without interrupting the airflow.

The four main components of all electrostatic precipitators are:

- Power supply unit, to provide high-voltage DC power
- Ionizing section, to impart a charge to particulates in the gas stream
- A means of removing the collected particulates
- A housing to enclose the precipitator zone

The following factors affect the efficiency of electrostatic precipitators:

- Larger collection-surface areas and lower gas-flow rates increase efficiency because of the increased time available for electrical activity to treat the dust particles.
- An increase in the dust-particle migration velocity to the collecting electrodes increases efficiency. The migration velocity can be increased by:
    - o Decreasing the gas viscosity
    - o Increasing the gas temperature
    - o Increasing the voltage field

Types of Precipitators

There are two main types of precipitators:

- High-voltage, single-stage - Single-stage precipitators combine an ionization and a collection step. They are commonly referred to as Cottrell precipitators.
- Low-voltage, two-stage - Two-stage precipitators use a similar principle; however, the ionizing section is followed by collection plates.

Described below is the high-voltage, single-stage precipitator, which is widely used in minerals processing operations. The low-voltage, two-stage precipitator is generally used for filtration in air-conditioning systems.

Plate Precipitators

The majority of electrostatic precipitators installed are the plate type. Particles are collected on flat, parallel surfaces that are 8 to 12 in. (20 to 30 cm) apart, with a series of discharge electrodes spaced along the centerline of two adjacent plates. The contaminated gases pass through the passage between the plates, and the particles become charged and adhere to the collection plates. Collected particles are usually removed by rapping the plates and deposited in bins or hoppers at the base of the precipitator.

Tubular Precipitators

Tubular precipitators consist of cylindrical collection electrodes with discharge electrodes located on the axis of the cylinder. The contaminated gases flow around the discharge electrode and up through the inside of the cylinders. The charged particles are collected on the grounded walls of the cylinder. The collected dust is removed from the bottom of the cylinder.

Tubular precipitators are often used for mist or fog collection or for adhesive, sticky, radioactive, or extremely toxic materials.

Unit Collectors

Unlike central collectors, unit collectors control contamination at its source. They are small and self-contained, consisting of a fan and some form of dust collector. They are suitable for isolated, portable, or frequently moved dust-producing operations, such as bins and silos or remote belt-conveyor transfer points. Advantages of unit collectors include small space requirements, the return of collected dust to main material flow, and low initial cost. However, their dust-holding and storage capacities, servicing facilities, and maintenance periods have been sacrificed.

A number of designs are available, with capacities ranging from 200 to 2,000 ft^3/min (90 to 900 L/s). There are two main types of unit collectors:

- Fabric collectors, with manual shaking or pulse-jet cleaning - normally used for fine dust

- Cyclone collectors - normally used for coarse dust

Fabric collectors are frequently used in minerals processing operations because they provide high collection efficiency and uninterrupted exhaust airflow between cleaning cycles. Cyclone collectors are used when coarser dust is generated, as in woodworking, metal grinding, or machining.

The following points should be considered when selecting a unit collector:

- Cleaning efficiency must comply with all applicable regulations.

- The unit maintains its rated capacity while accumulating large amounts of dust between cleanings.

- Simple cleaning operations do not increase the surrounding dust concentration.

- Has the ability to operate unattended for extended periods of time (for example, 8 hours).

- Automatic discharge or sufficient dust storage space to hold at least one week's accumulation.

- If renewable filters are used, they should not have to be replaced more than once a month.

- Durable.

- Quiet.

Use of unit collectors may not be appropriate if the dust-producing operations are located in an area where central exhaust systems would be practical. Dust removal and servicing requirements are expensive for many unit collectors and are more likely to be neglected than those for a single, large collector.

Selecting a Dust Collector

Dust collectors vary widely in design, operation, effectiveness, space requirements, construction, and capital, operating, and maintenance costs. Each type has advantages and disadvantages. However, the selection of a dust collector should be based on the following general factors:

- Dust concentration and particle size - For minerals processing operations, the dust concentration can range from 0.1 to 5.0 grains (0.32 g) of dust per cubic feet of air (0.23 to 11.44 grams per standard cubic meter), and the particle size can vary from 0.5 to 100 micrometres (μm) in diameter.

- Degree of dust collection required - The degree of dust collection required depends on its potential as a health hazard or public nuisance, the plant location, the allowable emission rate, the nature of the dust, its salvage value, and so forth. The selection of a collector should be based on the efficiency required and should consider the need for high-efficiency, high-cost equipment, such as electrostatic precipitators; high-efficiency, moderate-cost equipment, such as baghouses or wet scrubbers; or lower cost, primary units, such as dry centrifugal collectors.

- Characteristics of airstream - The characteristics of the airstream can have a significant impact on collector selection. For example, cotton fabric filters cannot be used where air temperatures exceed 180 °F (82 °C). Also, condensation of steam or water vapor can blind bags. Various chemicals can attack fabric or metal and cause corrosion in wet scrubbers.

- Characteristics of dust - Moderate to heavy concentrations of many dusts (such as dust from silica sand or metal ores) can be abrasive to dry centrifugal collectors. Hygroscopic material can blind bag collectors. Sticky material can adhere to collector elements and plug passages. Some particle sizes and shapes may rule out certain types of fabric collectors. The combustible nature of many fine materials rules out the use of electrostatic precipitators.

- Methods of disposal - Methods of dust removal and disposal vary with the material, plant process, volume, and type of collector used. Collectors can unload continuously or in batches. Dry materials can create secondary dust problems during unloading and disposal that do not occur with wet collectors. Disposal of wet slurry or sludge can be an additional material-handling problem; sewer or water pollution problems can result if wastewater is not treated properly.

Fan and Motor

The fan and motor system supplies mechanical energy to move contaminated air from the dust-producing source to a dust collector.

Types of Fans

There are two main kinds of industrial fans:

- Centrifugal fans

- Axial-flow fans

Centrifugal Fans

Centrifugal fans consist of a wheel or a rotor mounted on a shaft that rotates in a scroll-shaped housing. Air enters at the eye of the rotor, makes a right-angle turn, and is forced through the blades of the rotor by centrifugal force into the scroll-shaped housing. The centrifugal force imparts static pressure to the air. The diverging shape of the scroll also converts a portion of the velocity pressure into static pressure.

There are three main types of centrifugal fans:

- Radial-blade fans - Radial-blade fans are used for heavy dust loads. Their straight, radial blades do not get clogged with material, and they withstand considerable abrasion. These fans have medium tip speeds and medium noise factors.

- Backward-blade fans - Backward-blade fans operate at higher tip speeds and thus are more efficient. Since material may build up on the blades, these fans should be used after a dust collector. Although they are noisier than radial-blade fans, backward-blade fans are commonly used for large-volume dust collection systems because of their higher efficiency.

- Forward-curved-blade fans - These fans have curved blades that are tipped in the direction of rotation. They have low space requirements, low tip speeds, and a low noise factor. They are usually used against low to moderate static pressures.

Axial-flow Fans

Axial-flow fans are used in systems that have low resistance levels. These fans move the air parallel to the fan's axis of rotation. The screw-like action of the propellers moves the air in a straight-through parallel path, causing a helical flow pattern.

The three main kinds of axial fans are-

- Propeller fans - These fans are used to move large quantities of air against very low static pressures. They are usually used for general ventilation or dilution ventilation and are good in developing up to 0.5 in. wg (124.4 Pa).

- Tube-axial fans - Tube-axial fans are similar to propeller fans except they are mounted in a tube or cylinder. Therefore, they are more efficient than propeller fans and can develop up to 3 to 4 in. wg (743.3 to 995 Pa). They are best suited for moving air containing substances such as condensible fumes or pigments.

- Vane-axial fans - Vane-axial fans are similar to tube-axial fans except air-straightening vanes are installed on the suction or discharge side of the rotor. They are easily adapted to multistaging and can develop static pressures as high as 14 to 16 in. wg (3.483 to 3.98 kPa). They are normally used for clean air only.

Fan selection

When selecting a fan, the following points should be considered:

- Volume required

- Fan static pressure

- Type of material to be handled through the fan (For example, a radial-blade fan should be used with fibrous material or heavy dust loads, and nonsparking construction must be used with explosive or inflammable materials.)

- Type of drive arrangement, such as direct drive or belt drive

- Space requirements

- Noise levels

- Operating temperature (For example, sleeve bearings are suitable to 250 °F/121.1 °C; ball bearings to 550 °F/287.8 °C)

- Sufficient size to handle the required volume and pressure with minimum horsepower

- Need for special coatings or construction when operating in corrosive atmospheres

- Ability of fan to accommodate small changes in total pressure while maintaining the necessary air volume

- Need for an outlet damper to control airflow during cold starts (If necessary, the damper may be interlocked with the fan for a gradual start until steady-state conditions are reached.)

Fan Rating Tables

After the above information is collected, the actual selection of fan size and speed is usually made from a rating table published by the fan manufacturer. This table is known as a multirating table, and it shows the complete range of capacities for a particular size of fan.

Points to note:

- The multirating table shows the range of pressures and speeds possible within the limits of the fan's construction.

- A particular fan may be available in different construction classes (identified as class I through IV) relating to its capabilities and limits.

- For a given pressure, the highest mechanical efficiency is usually found in the middle third of the volume column.

- A fan operating at a given speed can have an infinite number of ratings (pressure and volume) along the length of its characteristic curve. However, when the fan is installed in a dust collection system, the point of rating can only be at the point at which the system resistance curve intersects the fan characteristic curve.

- In a given system, a fan at a fixed speed or at a fixed blade setting can have a single rating only. This rating can be changed only be changing the fan speed, blade setting, or the system resistance.

- For a given system, an increase in exhaust volume will result in increases in static and total pressures. For example, for a 20% increase in exhaust volume in a system with 5 in. pressure loss, the new pressure loss will be $5 \times (1.20)^2 = 7.2$ in.

- For rapid estimates of probable exhaust volumes available for a given motor size, the equation for brake horsepower, as illustrated, can be useful.

Fan installation Typical fan discharge conditions Fan ratings for volume and static pressure, as described in the multirating tables, are based on the tests conducted under ideal conditions. Often, field installation creates airflow problems that reduce the fan's air delivery. The following points should be considered when installing the fan:

- Avoid installation of elbows or bends at the fan discharge, which will lower fan performance by increasing the system's resistance.

- Avoid installing fittings that may cause non-uniform flow, such as an elbow, mitred elbow, or square duct.

- Check that the fan impeller is rotating in the proper direction-clockwise or counterclockwise.

- For belt-driven fans-

    o   Check that the motor sheave and fan sheave are aligned properly.

    o   Check for proper belt tension.

- Check the passages between inlets, impeller blades, and inside of housing for buildup of dirt, obstructions, or trapped foreign matter.

Electric Motors

Electric motors are used to supply the necessary energy to drive the fan.

Integral-horsepower electric motors are normally three-phase, alternating-current motors. Fractional-horsepower electric motors are normally single-phase, alternating-current motors and are used when less than 1 hp (0.75 kW) is required. Since most dust collection systems require motors with more than 1 hp (0.75 kW), only integral-horsepower motors are discussed here.

The two most common types of integral-horsepower motors used in dust collection systems are-

- Squirrel-cage motors - These motors have a constant speed and are of a nonsynchronous, induction type.

- Wound-rotor motors - These motors are also known as slip-ring motors. They are general-purpose or continuous-rated motors and are chiefly used when an adjustable-speed motor is desired.

Squirrel-cage and wound-rotor motors are further classified according to the type of enclosure they use to protect their interior windings. These enclosures fall into two broad categories:

- Open

- Totally enclosed

Drip-proof and splash-proof motors are open motors. They provide varying degrees of protection; however, they should not be used where the air contains substances that might be harmful to the interior of the motor.

Totally enclosed motors are weather-protected with the windings enclosed. These enclosures prevent free exchange of air between the inside and the outside, but they are not airtight.

Totally enclosed, fan-cooled (TEFC) motors are another kind of totally enclosed motor. These motors are the most commonly used motors in dust collection systems. They have an integral-cooling fan outside the enclosure, but within the protective shield, that directs air over the enclosure.

Both open and totally enclosed motors are available in explosion-proof and dust-ignition-proof models to protect against explosion and fire in hazardous environments.

Motors are selected to provide sufficient power to operate fans over the full range of process conditions (temperature and flow rate).

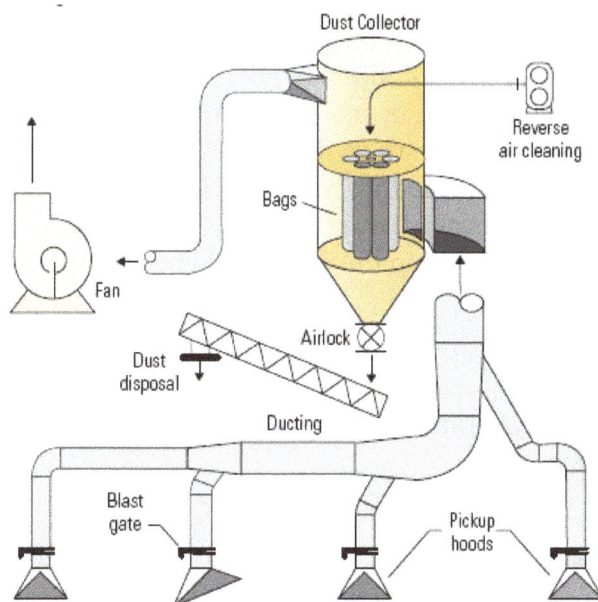

Figure 1. Dust Collection System Example

Configurations

Dust collectors can be configured into one of five common types.

1. Ambient units - Ambient units are free-hanging systems for use when applications limit the use of source-capture arms or ductwork.

2. Collection booths - Collector booths require no ductwork, and allow the worker greater freedom of movement. They are often portable.

3. Downdraft tables - A downdraft table is a self-contained portable filtration system that removes harmful particulates and returns filtered air back into the facility with no external ventilation required.

4. Source collector or Portable units - Portable units are for collecting dust, mist, fumes, or smoke at the source.

5. Stationary units - An example of a stationary collector is a baghouse.

Parameters Involved in Specifying Dust Collectors

Important parameters in specifying dust collectors include airflow the velocity of the air stream created by the vacuum producer; system power, the power of the system motor, usually specified in horsepower; storage capacity for dust and particles, and minimum particle size filtered by the unit. Other considerations when choosing a dust collection system include the temperature, moisture content, and the possibility of combustion of the dust being collected.

Systems for fine removal may only contain a single filtration system (such as a filter bag or cartridge). However, most units utilize a primary and secondary separation/filtration system. In many cases the heat or moisture content of dust can negatively affect the filter media of a baghouse or cartridge dust collector. A cyclone separator or dryer may be placed before these units to reduce heat or moisture content before reaching the filters. Furthermore, some units may have third and fourth stage filtration. All separation and filtration systems used within the unit should be specified.

A baghouse is an air pollution abatement device used to trap particulate by filtering gas streams through large fabric bags. They are typically made of glass fibers or fabric.

A cyclone separator is an apparatus for the separation, by centrifugal means, of fine particles suspended in air or gas.

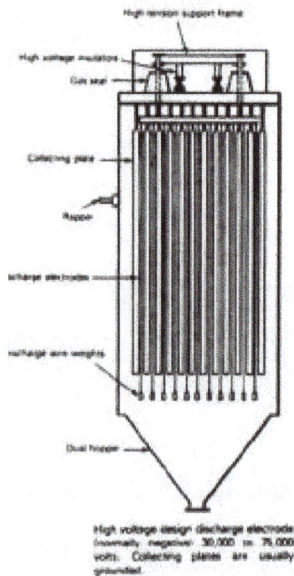

Electrostatic Precipitator

Electrostatic precipitators are a type of air cleaner, which charges particles of dust by passing dust-laden air through a strong (50-100 kV) electrostatic field. This causes the particles to be attracted to oppositely charged plates so that they can be removed from the air stream.

An impinger system is a device in which particles are removed by impacting the aerosol particles into a liquid. Modular media type units combine a variety of specific filter modules in one unit. These systems can provide solutions to many air contaminant problems. A typical system incorporates a series of disposable or cleanable pre-filters, a disposable vee-bag or cartridge filter. HEPA or carbon final filter modules can also be added. Various models are available, including free-hanging or ducted installations, vertical or horizontal mounting, and fixed or portable configurations. Filter cartridges are made out of a variety of synthetic fibers and are capable of collecting sub-micrometre particles without creating an excessive pressure drop in the system. Filter cartridges require periodic cleaning.

A wet scrubber, or venturi scrubber, is similar to a cyclone but it has an orifice unit that sprays water into the vortex in the cyclone section, collecting all of the dust in a slurry system. The water media can be recirculated and reused to continue to filter the air. Eventually the solids must be removed from the water stream and disposed of.

Filter Cleaning Methods

Online cleaning – automatically timed filter cleaning which allows for continuous, uninterrupted dust collector operation for heavy dust operations.

Offline cleaning – filter cleaning accomplished during dust collector shut down. Practical whenever the dust loading in each dust collector cycle does not exceed the filter capacity. Allows for maximum effectiveness in dislodging and disposing of dust.

On-demand cleaning – filter cleaning initiated automatically when the filter is fully loaded, as determined by a specified drop in pressure across the media surface.

Reverse-pulse/Reverse-jet cleaning – Filter cleaning method which delivers blasts of compressed air from the clean side of the filter to dislodge the accumulated dust cake.

Impact/Rapper cleaning – Filter cleaning method in which high-velocity compressed air forced through a flexible tube results in a random rapping of the filter to dislodge the dust cake. Especially effective when the dust is extremely fine or sticky.

Scrubber

Scrubber systems are a diverse group of air pollution control devices that can be used to remove some particulates and/or gases from industrial exhaust streams. The first air scrubber was designed to remove carbon dioxide from the air of an early submarine, the Ictineo I, a role for which they continue to be used till today. Traditionally, the term "scrubber" has referred to pollution control devices that use liquid to wash unwanted pollutants from a gas stream. Recently, the term has also been used to describe systems that inject a dry reagent or slurry into a dirty exhaust stream to "wash out" acid gases. Scrubbers are one of the primary devices that control gaseous emissions, especially acid gases. Scrubbers can also be used for heat recovery from hot gases by flue-gas condensation.

There are several methods to remove toxic or corrosive compounds from exhaust gas and neutralize it.

Combustion

Combustion is sometimes the cause of harmful exhausts, but, in many cases, combustion may also be used for exhaust gas cleaning if the temperature is high enough and enough oxygen is available.

Wet Scrubbing

The exhaust gases of combustion may contain substances considered harmful to the environment, and the scrubber may remove or neutralize those. A wet scrubber is used for cleaning air, fuel gas or other gases of various pollutants and dust particles. Wet scrubbing works via the contact of target compounds or particulate matter with the scrubbing solution. Solutions may simply be water (for dust) or solutions of reagents that specifically target certain compounds.

Process exhaust gas can also contain water-soluble toxic and/or corrosive gases like hydrochloric acid (HCl) or ammonia (NH_3). These can be removed very well by a wet scrubber.

Removal efficiency of pollutants is improved by increasing residence time in the scrubber or by the increase of surface area of the scrubber solution by the use of a spray nozzle, packed towers or an aspirator. Wet scrubbers may increase the proportion of water in the gas, resulting in a visible stack plume, if the gas is sent to a stack.

Wet scrubbers can also be used for heat recovery from hot gases by flue-gas condensation. In this mode, termed a condensing scrubber, water from the scrubber drain is circulated through a cooler to the nozzles at the top of the scrubber. The hot gas enters the scrubber at the bottom. If the gas temperature is above the water dew point, it is initially cooled by evaporation of water drops. Further cooling cause water vapors to condense, adding to the amount of circulating water.

The condensation of water release significant amounts of low temperature heat (more than 2 gigajoules (560 kWh) per ton of water), that can be recovered by the cooler for e.g. district heating purposes.

Excess condensed water must continuously be removed from the circulating water.

The gas leaves the scrubber at its dew point, so even though significant amounts of water may have been removed from the cooled gas, it is likely to leave a visible stack plume of water vapor.

Dry Scrubbing

A dry or semi-dry scrubbing system, unlike the wet scrubber, does not saturate the flue gas stream that is being treated with moisture. In some cases no moisture is added, while in others only the amount of moisture that can be evaporated in the flue gas without condensing is added. Therefore, dry scrubbers generally do not have a stack steam plume or wastewater handling/disposal requirements. Dry scrubbing systems are used to remove acid gases (such as SO_2 and HCl) primarily from combustion sources.

There are a number of dry type scrubbing system designs. However, all consist of two main sections or devices: a device to introduce the acid gas sorbent material into the gas stream and a particulate matter control device to remove reaction products, excess sorbent material as well as any particulate matter already in the flue gas.

Dry scrubbing systems can be categorized as dry sorbent injectors (DSIs) or as spray dryer absorbers (SDAs). Spray dryer absorbers are also called semi-dry scrubbers or spray dryers.

Dry scrubbing systems are often used for the removal of odorous and corrosive gases from wastewater treatment plant operations. The medium used is typically an activated alumina compound impregnated with materials to handle specific gases such as hydrogen sulfide. Media used can be mixed together to offer a wide range of removal for other odorous compounds such as methyl mercaptans, aldehydes, volatile organic compounds, dimethyl sulfide, and dimethyl disulfide.

Dry sorbent injection involves the addition of an alkaline material (usually hydrated lime, soda ash, or sodium bicarbonate) into the gas stream to react with the acid gases. The sorbent can be injected directly into several different locations: the combustion process, the flue gas duct (ahead of the particulate control device), or an open reaction chamber (if one exists). The acid gases react with the alkaline sorbents to form solid salts which are removed in the particulate control device. These simple systems can achieve only limited acid gas (SO_2 and HCl) removal efficiencies. Higher collection efficiencies can be achieved by increasing the flue gas humidity (i.e., cooling using water spray). These devices have been used on medical waste incinerators and a few municipal waste combustors.

In spray dryer absorbers, the flue gases are introduced into an absorbing tower (dryer) where the gases are contacted with a finely atomized alkaline slurry. Acid gases are absorbed by the slurry mixture and react to form solid salts which are removed by the particulate control device. The heat of the flue gas is used to evaporate all the water droplets, leaving a non-saturated flue gas to exit the absorber tower. Spray dryers are capable of achieving high (80+%) acid gas removal efficiencies. These devices have been used on industrial and utility boilers and municipal waste incinerators.

Absorber

Many chemicals can be removed from exhaust gas also by using absorber material. The flue gas is passed through a cartridge which is filled with one or several absorber materials and has been adapted to the chemical properties of the components to be removed. This type of scrubber is sometimes also called dry scrubber. The absorber material has to be replaced after its surface is saturated.

Mercury Removal

Mercury is a highly toxic element commonly found in coal and municipal waste. Wet scrubbers are only effective for removal of soluble mercury species, such as oxidized mercury, Hg^{2+}. Mercury vapor in its elemental form, Hg^o, is insoluble in the scrubber slurry and not removed. Therefore, an additional process of Hg^o conversion is required to complete mercury capture. Usually halogens are added to the flue gas for this purpose. The type of coal burned as well as the presence of a selective catalytic reduction unit both affect the ratio of elemental to oxidized mercury in the flue gas and thus the degree to which the mercury is removed.

In July 2015, one study found that some mercury scrubbers installed on coal power plants inadvertently capture PAH (polycyclic aromatic hydrocarbons) emissions as well.

Scrubber Waste Products

One side effect of scrubbing is that the process only moves the unwanted substance from the exhaust gases into a liquid solution, solid paste or powder form. This must be disposed of safely, if it can not be reused.

For example, mercury removal results in a waste product that either needs further processing to extract the raw mercury, or must be buried in a special hazardous wastes landfill that prevents the mercury from seeping out into the environment.

As an example of reuse, limestone-based scrubbers in coal-fired power plants can produce a synthetic gypsum of sufficient quality that can be used to manufacture drywall and other industrial products.

Bacteria Spread

Poorly maintained scrubbers have the potential to spread disease-causing bacteria. The problem is a result of inadequate cleaning. For example, the cause of a 2005 outbreak of Legionnaires' disease in Norway was just a few infected scrubbers. The outbreak caused 10 deaths and more than 50 cases of infection.

Wet Scrubber

The term wet scrubber describes a variety of devices that remove pollutants from a furnace flue gas or from other gas streams. In a wet scrubber, the polluted gas stream is brought into contact with the scrubbing liquid, by spraying it with the liquid, by forcing it through a pool of liquid, or by some other contact method, so as to remove the pollutants.

Design

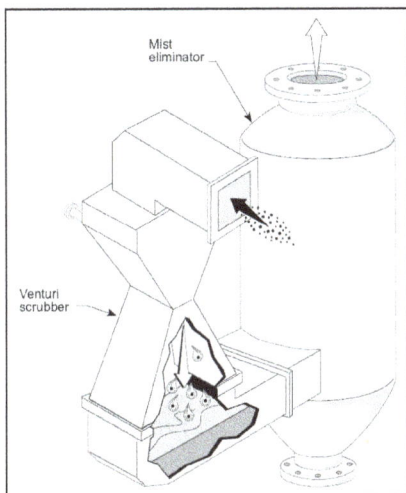

A venturi scrubber design. The mist eliminator for a venturi scrubber is often a separate device called a cyclonic separator

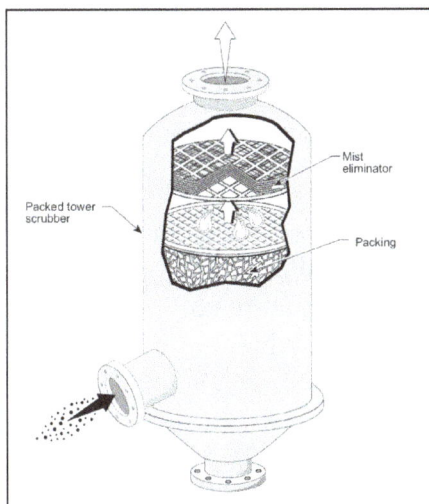

A packed bed tower design where the mist eliminator is built into the top of the structure. Various tower designs exist

The design of wet scrubbers or any air pollution control device depends on the industrial process conditions and the nature of the air pollutants involved. Inlet gas characteristics and dust properties (if particles are present) are of primary importance. Scrubbers can be designed to collect particulate matter and/or gaseous pollutants. The versatility of wet scrubbers allow them to be built in numerous configurations, all designed to provide good contact between the liquid and polluted gas stream.

Wet scrubbers remove dust particles by *capturing* them in liquid droplets. The droplets are then collected, the liquid *dissolving* or *absorbing* the pollutant gases. Any droplets that are in the scrubber inlet gas must be separated from the outlet gas stream by means of another device referred to as a mist eliminator or entrainment separator (these terms are interchangeable). Also, the resultant scrubbing liquid must be treated prior to any ultimate discharge or being reused in the plant.

A wet scrubber's ability to collect small particles is often directly proportional to the power input into the scrubber. Low energy devices such as spray towers are used to collect particles larger than 5 micrometers. To obtain high efficiency removal of 1 micrometer (or less) particles generally requires high energy devices such as venturi scrubbers or augmented devices such as condensation scrubbers. Additionally, a properly designed and operated entrainment separator or mist eliminator is important to achieve high removal efficiencies. The greater the number of liquid droplets that are not captured by the mist eliminator, the higher the potential emission levels.

Wet scrubbers that remove gaseous pollutants are referred to as *absorbers*. Good gas-to-liquid contact is essential to obtain high removal efficiencies in absorbers. A number of wet scrubber designs are used to remove gaseous pollutants, with the packed tower and the plate tower being the most common.

If the gas stream contains both particle matter and gases, wet scrubbers are generally the only single air pollution control device that can remove both pollutants. Wet scrubbers can achieve high removal efficiencies for either particles or gases and, in some instances, can achieve a high removal efficiency for both pollutants in the same system. However, in many cases, the best operating conditions for particles collection are the poorest for gas removal.

In general, obtaining high simultaneous gas and particulate removal efficiencies requires that one of them be easily collected (i.e., that the gases are very soluble in the liquid or that the particles are large and readily captured) or by the use of a scrubbing reagent such as lime or sodium hydroxide.

Advantages and Disadvantages

For particulate control, wet scrubbers (also referred to as wet collectors) are evaluated against fabric filters and electrostatic precipitators (ESPs). Some *advantages* of wet scrubbers over these devices are as follows:

- Wet scrubbers have the ability to handle high temperatures and moisture.

- In wet scrubbers, flue gases are cooled, resulting in smaller overall size of equipment.

- Wet scrubbers can remove both gases and particulate matter.

- Wet scrubbers can neutralize corrosive gases.

Some *disadvantages* of wet scrubbers include corrosion, the need for entrainment separation or mist removal to obtain high efficiencies and the need for treatment or reuse of spent liquid.

Wet scrubbers have been used in a variety of industries such as acid plants, fertilizer plants, steel mills, asphalt plants, and large power plants.

Relative advantages and disadvantages of wet scrubbers compared to other control devices	
Advantages	**Disadvantages**
• Small space requirements: Scrubbers reduce the temperature and volume of the unsaturated exhaust stream. Therefore, vessel sizes, including fans and ducts downstream, are smaller than those of other control devices. Smaller sizes result in lower capital costs and more flexibility in site location of the scrubber. • No secondary dust sources: Once particulate matter is collected, it cannot escape from hoppers or during transport. • Handles high-temperature, high-humidity gas streams: No temperature limits or condensation problems can occur as in baghouses or ESPs. • Minimal fire and explosion hazards: Various dry dusts are flammable. Using water eliminates the possibility of explosions. • Ability to collect both gases and particulate matter.	• Corrosion problems: Water and dissolved pollutants can form highly corrosive acid solutions. Proper construction materials are very important. Also, wet-dry interface areas can result in corrosion. • High power requirements: High collection efficiencies for particulate matter are attainable only at high pressure drops, resulting in high operating costs. • Water-disposal problems: Settling ponds or sludge clarifiers may be needed to meet waste-water regulations. • Difficult product recovery: Dewatering and drying of scrubber sludge make recovery of any dust for reuse very expensive and difficult.

Components

Wet scrubber systems generally consist of the following components:

- Ductwork and fan system

- A saturation chamber (optional)

- Scrubbing vessel

- Entrainment separator or mist eliminator

- Pumping (and possible recycle system)

- Spent scrubbing liquid treatment and/or reuse system

- An exhaust stack

A typical wet scrubbing process can be described as follows:

- Hot flue gas from a furnace enters a saturator (if present) where gases are cooled and humidified prior to entering the scrubbing area. The saturator removes a small percentage of the particulate matter present in the flue gas.

- Next, the gas enters a venturi scrubber where approximately half of the gases are removed. Venturi scrubbers have a minimum particle removal efficiency of 95%.

- The gas flows through a second scrubber, a packed bed absorber, where the rest of the gases (and particulate matter) are collected.

- An entrainment separator or mist eliminator removes any liquid droplets that may have become entrained in the flue gas.

- A recirculation pump moves some of the spent scrubbing liquid back to the venturi scrubber where it is recycled and the remainder is sent to a treatment system.

- Treated scrubbing liquid is recycled back to the saturator and the packed bed absorber.

- Fans and ductwork move the flue gas stream through the system and eventually out the stack.

Categorization

Since wet scrubbers vary greatly in complexity and method of operation, devising categories into which all of them neatly fit is extremely difficult. Scrubbers for particle collection are usually categorized by the gas-side pressure drop of the system. Gas-side pressure drop refers to the pressure difference, or pressure drop, that occurs as the exhaust gas is pushed or pulled through the scrubber, disregarding the pressure that would be used for pumping or spraying the liquid into the scrubber.

Scrubbers may be classified *by pressure drop* as follows:

- *Low-energy scrubbers* have pressure drops of less than 12.7 cm (5 in) of water.

- *Medium-energy scrubbers* have pressure drops between 12.7 and 38.1 cm (5 and 15 in) of water.

- *High-energy scrubbers* have pressure drops greater than 37.1 cm (15 in) of water.

However, most scrubbers operate over a wide range of pressure drops, depending on their specific application, thereby making this type of categorization difficult.

Another way to classify wet scrubbers is by their *use* - to primarily collect either particulates or gaseous pollutants. Again, this distinction is not always clear since scrubbers can often be used to remove both types of pollutants.

Wet scrubbers can also be categorized by the manner in which the gas and liquid phases are brought into contact. Scrubbers are designed to use power, or energy, from the gas stream or the liquid stream, or some other method to bring the pollutant gas stream into contact with the liquid. These categories are given in Table 2.

Categories of wet collectors by energy source used for contact	
Wet collector	**Energy source used for gas-liquid contact**
• Gas-phase contacting	• Gas stream
• Liquid-phase contacting	• Liquid stream
• Wet film	• Liquid and gas streams
• Combination	• Energy source:
o Liquid phase and gas phase	o Liquid and gas streams
o Mechanically aided	o Mechanically driven rotor

Material of Construction and Design

Corrosion can be a prime problem associated with chemical industry scrubbing systems. Fibre-reinforced plastic and dual keys are often used as most dependable materials of construction.

Baffle Spray Scrubber

Baffle spray scrubbers are a technology for air pollution control. They are very similar to spray towers in design and operation. However, in addition to using the energy provided by the spray nozzles, baffles are added to allow the gas stream to atomize some liquid as it passes over them.

A simple baffle scrubber system is shown in Figure 1. Liquid sprays capture pollutants and also remove collected particles from the baffles. Adding baffles slightly increases the pressure drop of the system.

This type of technology is a part of the group of air pollution controls collectively referred to as wet scrubbers.

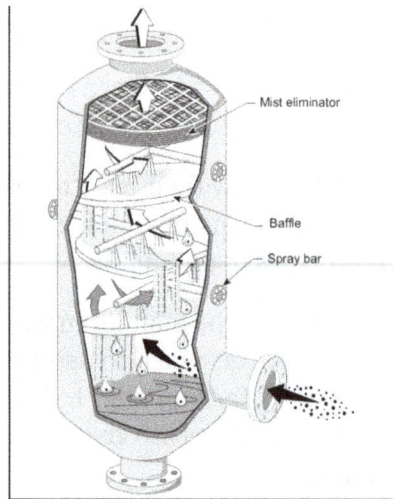

Figure 1 - Baffle spray scrubber

A number of wet-scrubber designs use energy from both the gas stream and liquid stream to collect pollutants. Many of these combination devices are available commercially.

A seemingly unending number of scrubber designs have been developed by changing system geometry and incorporating vanes, nozzles, and baffles.

Particle Collection

These devices are used much the same as spray towers - to preclean or remove particles larger than 10 µm in diameter. However, they will tend to plug or corrode if particle concentration of the exhaust gas stream is high.

Gas Collection

Even though these devices are not specifically used for gas collection, they are capable of a small amount of gas absorption because of their large wetted surface.

Summary

These devices are most commonly used as precleaners to remove large particles (>10 µm in diameter). The pressure drops across baffle scrubbers are usually low, but so are the collection efficiencies. Maintenance problems are minimal. The main problem is the buildup of solids on the baffles.

Ejector Venturi Scrubber

An ejector or venturi scrubber is an industrial pollution control device, usually installed on the exhaust flue gas stacks of large furnaces, but may also be used on any number of other air exhaust systems. This type of technology is a part of the group of air pollution controls collectively referred to as wet scrubbers.

Like a spray tower an ejector venturi scrubber uses a preformed spray. However, in an ejector venturi scrubber only a single nozzle is used instead of many nozzles. This nozzle operates at higher pressures and higher injection rates than those in most spray chambers. The high-pressure spray nozzle (up to 689 kPa or 100 psig) is aimed at the throat section of a venturi constriction.

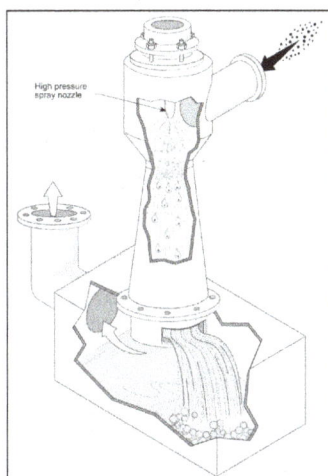

High pressure
spray nozzle

Figure 1 - Ejector venturi scrubber

The ejector venturi is unique among available scrubbing systems since it can move the process gas without the aid of a fan or blower. The liquid spray coming from the nozzle creates a partial vacuum in the side duct of the scrubber. The partial vacuum is due to the Bernoulli effect, and is similar to water aspirators used in chemistry labs. This partial vacuum can be used to move the process gas through the venturi as well as through the facility's process system. In the case of explosive or extremely corrosive atmospheres, the elimination of a fan in the system can avoid many potential problems.

The energy for the formation of scrubbing droplets comes from the injected liquid. The high pressure sprays passing through the venturi throat form numerous fine liquid droplets that provide turbulent mixing between the gas and liquid phases. Very high liquid-injection rates are used to provide the gas-moving capability and higher collection efficiencies. As with other types of venturis, a means of separating entrained liquid from the gas stream must be installed. Entrainment separators are commonly used to remove remaining small droplets.

Particle Collection

Ejector venturis are effective in removing particles larger than 1.0 μm in diameter. These scrubbers are not used on submicrometer-sized particles unless the particles are condensable [Gilbert, 1977]. Particle collection occurs primarily by impaction as the exhaust gas (from the process) passes through the spray.

The turbulence that occurs in the throat area also causes the particles to contact the wet droplets and be collected. Particle collection efficiency increases with an increase in nozzle pressure and/or an increase in the liquid-to-gas ratio. Increases in either of these two operating parameters will also result in an increase in pressure drop for a given system. Therefore, an increase in pressure drop also increases particle collection efficiency. Ejector venturis operate at higher L/G ratios than most other particulate scrubbers (i.e., 7 to 13 l/m³ compared to 0.4-2.7 l/m³ for most other designs).

Gas Collection

Ejector venturis have a short gas-liquid contact time because the exhaust gas velocities through the vessel are very high. This short contact time limits the absorption efficiency of the system. Although ejector venturis are not used primarily for gas removal, they can be effective if the gas is very soluble or if a very reactive scrubbing reagent is used. In these instances, removal efficiencies of as high as 95% can be achieved [Gilbert, 1977].

Maintenance Problems

Ejector venturis are subject to abrasion problems in the high-velocity areas - nozzle and throat. Both must be constructed of wear-resistant materials because of the high liquid injection rates and nozzle pressures. Maintaining the pump that recirculates liquid is also very important. In addition, the high gas velocities necessitate the use of entrainment separators to prevent excessive liquid carryover. The separators should be easily accessible or removable so that they can be cleaned if plugging occurs.

Summary

Because of their open design and the fact that they do not require a fan, ejector venturis are capable of handling a wide range of corrosive and/or sticky particles. However, they are not very effective in removing submicrometer particles. They have an advantage in being able to handle small, medium and large exhaust flows. They can be used singly or in multiple stages of two or more in series, depending on the specific application. Multiple-stage systems have been used where extremely high collection efficiency of particles or gaseous pollutants was necessary. Multiple-stage systems provide increased gas-liquid contact time, thus increasing absorption efficiency.

LO-NOx Burner

A LO NOx burner is a type of burner that is typically used in utility boilers to produce steam and electricity.

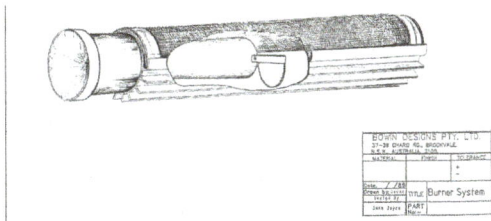

One of John Joyce's early sketches of the Low NOx burner

Background

John Joyce the inventor of the LO-NOx burner at Australian
Gas Association Conference in the early 1990s

The First Discovery

Around 1986 John Joyce (of Bowin Cars fame), an influential Australian inventor, first learned about oxides of nitrogen (NO_x) and their role in the production of smog and acid rain. His first introduction to the complexities of the subject was brought about by the work of Fred Barnes and Dr John Bromley from the state Energy Commission of Western Australia.

The vast majority of the research and development stretching back over twenty years was about large scale industrial burners and complex mechanisms which, in the end, did not produce what one would consider low NO_x (2 ng/J or ~ 4 ppm at 0% O_2 on dry basis).

In fact at that time, 15 ng/J NO_2 appears to have been considered low NO_2. The one clear message that did flow through all the mass of information he studied, was the effect of temperature on the formation of NO_x.

"Need is the Mother of Invention"

In the late 1980s, Health and Environment Authorities in Australia raised concerns about the indoor air quality and the extent that particularly older style unflued gas heaters were contributing to higher than acceptable levels of nitrogen dioxide (NO_2).

Consequently in 1989 the New South Wales Department of School Education initiated an extensive investigation of nitrogen dioxide in schools throughout New South Wales. As an interim measure the Health Authorities advised that a level of 0.3 ppm NO_2 should become the upper limit for classrooms. The Australian Gas Association in turn reduced the indoor emission rate of NO_2 for unflued gas heaters from 15 to 5 ng/J and this remains the current limit. The New South Wales government, through the Public Works Department, also re-evaluated alternative methods of heating classrooms, to ensure a safe and healthy environment for students.

It was in this context, that John Joyce's company Bowin Technology embarked on a major research & development program aimed at minimising nitrogen dioxide emissions from unflued gas heaters. Bowin Technology set itself the task of solving the emission problem at its source: the gas burner. This was despite a generally long held belief by gas experts, that commercially warranted gas burner improvements could not deliver drastic nitrogen oxides (NO_x) reductions.

In 1989, an immediate call to reduce the indoor nitrogen dioxide (NO_2) level, was triggered by widely publicised articles and media coverage in New South Wales, highlighting the effect this chemical has on respiratory sensitive people, such as asthmatics and those with bronchial problems.

In the heat of the indoor air quality debate various State institutions in Australia were advised to switch to flued gas heaters and electric heating.

New South Wales in contrast, through combined action by Australian Gas Light Company, Health Authorities and the New South Wales Public Works Department, formulated initial indoor air quality guidelines. These guidelines formed the basis for Australian Gas Appliance Code restrictions for nitrogen dioxide NO_2 emissions from unflued heaters, now adopted Australia wide.

John Joyce became aware that no other overseas regulatory body made a distinction between NO and NO_2 in their environmental guidelines or codes. Furthermore it appeared that total nitrogen oxides level requirements were in place irrespective of whether emissions were flued or not.

Consequently John Joyce learned that a 'harmless' part of NO_x emissions, nitric oxide (NO), in the presence of hydrocarbons (such as household aerosol propellants, possible gas leaks and ingress of vehicle exhaust fume), converts to NO_2. This was found to be the case in the New South Wales school investigation. In a scientific sense it had become practice to calculate both NO + NO_2, when measuring oxides of nitrogen levels in emissions. Hence the now commonly used reference to "total NO_x".

Greenhouse Gas and Photochemical Smog

Natural gas by composition has a distinct advantage over other fossil fuels in terms of carbon dioxide, particulate and sulfur dioxides produced when converting to useful energy. In the early 1990s numerous countries were in the process of substituting oil and coal with natural gas for their energy and electric power needs.

To maintain this advantage as an "environmentally friendly" fuel, Australian gas utilities are effectively reducing gas losses (methane emissions) in their deliveries, and impose strict codes on appliance manufacturers and installers against gas leakage.

Nevertheless environmental experts see the production of oxides of nitrogen as a major menace in the formation of greenhouse gases and photochemical smog. The interaction of NOx with hydrocarbons from vehicle exhausts and sunlight, can also form low level ozone. In the stratosphere (some 25 km up). Ozone is helpful by absorbing the fiercer part of the ultraviolet radiation of the sun, but at ground level it damages materials and vegetation. It irritates throat, lungs and eyes, and strenuous exercise or work can become painful. Furthermore, the effectiveness of nitrous oxide as a greenhouse gas is magnified by its longer life than carbon dioxide, methane and CFC's.

In essence the rate at which low level ozone is formed is determined by hydrocarbons, whilst the availability of oxides of nitrogen influences the amount it produces. At this point the environmental debate takes a surprising turn as individual industries tend to blame each other's emission as a probable cause.

Best Available Control Technology (BACT)

It is well established that conventional "blue flame" or bunsen gas burners produce oxides of nitrogen at levels of 30-50 nanograms per joule and are as such not considered to have potential for NOx reduction. Surface combustion burners or radiant tile burners in comparison produce nitrogen oxides' levels 60-70% less. Therefore John Joyce's research into low NO_x burners revolved primarily around surface combustion techniques. Another issue was the effect combustion temperatures have on the formation of NO_x.

John Joyce's task became even more challenging when he decided not to direct his development towards radiant type surface combustion tiles. The use of radiant heating for most institutional purposes (other than spot heating) is considered impractical as is too hot close to the heater, while the loss of radiant heat over a distance to be reached is quite dramatic.

Investigations into numerous developments of other types of "low NOx" burners showed that so far such burners were either too complex in design or operation, too expensive or unsuitable. John Joyce's plan was to use high temperature steel mesh, and went on to produce scores of prototype burners until one showed "potential".

The scientific innovative nature of John Joyce's LO-NO_x technologies are confirmed by full patent protection in Australia, United States, United Kingdom, Japan, Italy and France.

In 1993 John Joyce received an Australian Design Award and Powerhouse Museum Selection status for his "SLE" heater range, which incorporate LO-NO$_x$ burners.

The Australian Academy of Design selected the SLE unflued gas heater range to be featured in the Design Showcase during the "Innovation by Design" National Conference in October 1994

In the United States, John Joyce's LO-NO$_x$ water heater burners have successfully undergone a series of exhaustive tests to prove that these particular burners do not act as an ignition source in the presence of flammable vapours, resulting from accidental fuel spillage. There have also been extensive tests carried out to verify its reduction of NO$_2$.

Energy Efficiencies

More tangible cost savings are defined when comparing the energy efficiencies of gas heaters with low NO$_x$ emissions with conventional flued types. Gas heaters with emission problems are flued and inherently lose substantial energies in the form of hot flue gases to the atmosphere. In addition, the choice of placement of flued heaters is greatly impaired due to flue installation restrictions.

In contrast, dedicated low emission gas heaters do not require a flue system. Furthermore, with the introduction of oxygen depletion sensors and thermostatic controls, they do not place critical reliance on ventilation as had been the case. These heaters can be positioned more conveniently and centralised to affect optimum warm air distribution. By definition unflued low NO$_x$ gas heaters are 100% efficient as all heat energy released from the flame is converted to useful heat.

Applications of Technology

- Gas-fired unflued gas heaters

- Gas-fired flued gas heaters

- Gas-fired storage water heaters

Catalytic Converter

A catalytic converter is an emissions control device that converts toxic gases and pollutants in exhaust gas to less toxic pollutants by catalyzing a redox reaction (an oxidation and a reduction reaction). Catalytic converters are used with internal combustion engines fueled by either petrol (gasoline) or diesel—including lean-burn engines as well as kerosene heaters and stoves.

A three-way catalytic converter on a gasoline-powered 1996 Dodge Ram Van

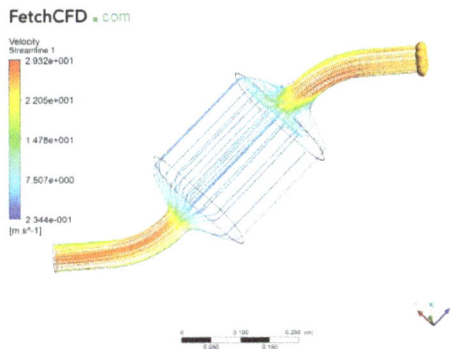

Simulation of flow inside a catalytic converter

The first widespread introduction of catalytic converters was in the United States automobile market. To comply with the U.S. Environmental Protection Agency's stricter regulation of exhaust emissions, most gasoline-powered vehicles starting with the 1975 model year must be equipped with catalytic converters. These "two-way" converters combined oxygen with carbon monoxide (CO) and unburned hydrocarbons (HC) to produce carbon dioxide (CO_2) and water (H_2O). In 1981, two-way catalytic converters were rendered obsolete by "three-way" converters that also reduce oxides of nitrogen (NOx); however, two-way converters are still used for lean-burn engines. This is because three-way-converters require either rich or stoichiometric combustion to successfully reduce NOx.

Although catalytic converters are most commonly applied to exhaust systems in automobiles, they are also used on electrical generators, forklifts, mining equipment, trucks, buses, locomotives and motorcycles. They are also used on some wood stoves to control emissions. This is usually in response to government regulation, either through direct environmental regulation or through health and safety regulations.

History

The catalytic converter was invented by Eugene Houdry, a French mechanical engineer and expert in catalytic oil refining, who moved to the United States in 1930. When the results of early studies of smog in Los Angeles were published, Houdry became concerned about the role of smoke stack exhaust and automobile exhaust in air pollution and founded a company called Oxy-Catalyst. Houdry first developed catalytic converters for smoke stacks called "cats" for short, and later developed catalytic converters for warehouse forklifts that used low grade, unleaded gasoline. In the mid-1950s, he began research to develop catalytic converters for gasoline engines used on cars. He was awarded United States Patent 2,742,437 for his work.

Widespread adoption of catalytic converters did not occur until more stringent emission control regulations forced the removal of the anti-knock agent tetraethyl lead from most types of gasoline. Lead is a "catalyst poison" and would effectively disable a catalytic converter by forming a coating on the catalyst's surface.

Catalytic converters were further developed by a series of engineers including John J. Mooney, Carl D. Keith, Antonio Eleazar at the Engelhard Corporation, creating the first production catalytic converter in 1973.

William C. Pfefferle developed a catalytic combustor for gas turbines in the early 1970s, allowing combustion without significant formation of nitrogen oxides and carbon monoxide.

Construction

Cutaway of a metal-core converter

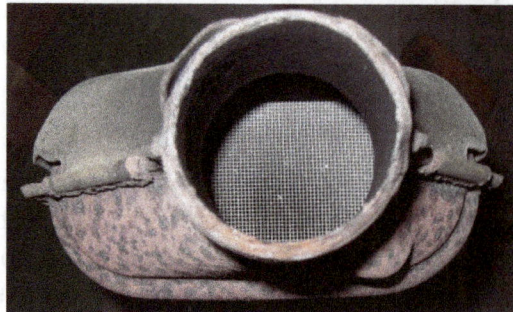

Ceramic-core converter

The catalytic converter's construction is as follows:

1. The catalyst support or substrate. For automotive catalytic converters, the core is usually a ceramic monolith with a honeycomb structure. Metallic foil monoliths made of Kanthal (FeCrAl) are used in applications where particularly high heat resistance is required. Either material is designed to provide a large surface area. The cordierite ceramic substrate used in most catalytic converters was invented by Rodney Bagley, Irwin Lachman and Ronald Lewis at Corning Glass, for which they were inducted into the National Inventors Hall of Fame in 2002.

2. The washcoat. A washcoat is a carrier for the catalytic materials and is used to disperse the materials over a large surface area. Aluminum oxide, titanium dioxide, silicon dioxide, or a mixture of silica and alumina can be used. The catalytic materials are suspended in the washcoat prior to applying to the core. Washcoat materials are selected to form a rough, irregular surface, which greatly increases the surface area compared to the smooth surface of the bare substrate. This in turn maximizes the catalytically active surface available to react with the engine exhaust. The coat must retain its surface area and prevent sintering of the catalytic metal particles even at high temperatures (1000 °C).

3. Ceria or ceria-zirconia. These oxides are mainly added as oxygen storage promoters.

4. The catalyst itself is most often a mix of precious metals. Platinum is the most active catalyst and is widely used, but is not suitable for all applications because of unwanted additional reactions and high cost. Palladium and rhodium are two other precious metals used. Rhodium is used as a reduction catalyst, palladium is used as an oxidation catalyst, and platinum is used both for reduction and oxidation. Cerium, iron, manganese and nickel are also used, although each has limitations. Nickel is not legal for use in the European Union because of its reaction with carbon monoxide into toxic nickel tetracarbonyl. Copper can be used everywhere except Japan.

Upon failure, a catalytic converter can be recycled into scrap. The precious metals inside the converter, including platinum, palladium and rhodium, are extracted.

Placement Of Catalytic Converters

Catalytic converters require temperature of 800 degrees Fahrenheit to efficiently convert harmful exhaust gases into inert ones, such as carbon dioxide and water vapor. So, first catalytic converters were placed close to the engine to ensure fast heating. However, such placing caused several problems, such as vapor lock.

As an alternative, catalytic converters were moved to a third of the way back from the engine, and were then placed underneath the vehicle.

In 1990s, integrated catalytic converters were developed, which, as the name suggests, were integrated into exhaust manifold assemblies. Their high efficiency, safety and space-saving capability quickly earned them a popularity. Today, almost every new vehicle sold in the United States is equipped with integrated catalytic converters.

Types

Two-way

A 2-way (or "oxidation", sometimes called an "oxi-cat") catalytic converter has two simultaneous tasks:

1. Oxidation of carbon monoxide to carbon dioxide: $2CO + O_2 \rightarrow 2CO_2$

2. Oxidation of hydrocarbons (unburned and partially burned fuel) to carbon dioxide and water: $C_xH_{2x+2} + [(3x+1)/2]\ O_2 \rightarrow xCO_2 + (x+1)\ H_2O$ (a combustion reaction)

This type of catalytic converter is widely used on diesel engines to reduce hydrocarbon and carbon monoxide emissions. They were also used on gasoline engines in American- and Canadian-market automobiles until 1981. Because of their inability to control oxides of nitrogen, they were superseded by three-way converters.

Three-way

Three-way catalytic converters (TWC) have the additional advantage of controlling the emission of nitric oxide and nitrogen dioxide, which are precursors to acid rain and smog.

Since 1981, "three-way" (oxidation-reduction) catalytic converters have been used in vehicle emission control systems in the United States and Canada; many other countries have also adopted stringent vehicle emission regulations that in effect require three-way converters on gasoline-powered vehicles. The reduction and oxidation catalysts are typically contained in a common housing; however, in some instances, they may be housed separately. A three-way catalytic converter has three simultaneous tasks:

1. Reduction of nitrogen oxides to nitrogen and oxygen: $2NO_x \rightarrow xO_2 + N_2$

2. Oxidation of carbon monoxide to carbon dioxide: $2CO + O_2 \rightarrow 2CO_2$

3. Oxidation of unburnt hydrocarbons (HC) to carbon dioxide and water: $C_xH_{2x+2} + [(3x+1)/2]O_2 \rightarrow xCO_2 + (x+1)H_2O$.

These three reactions occur most efficiently when the catalytic converter receives exhaust from an engine running slightly above the stoichiometric point. For gasoline combustion, this ratio is between 14.6 and 14.8 parts air to one part fuel, by weight. The ratio for Autogas (or liquefied petroleum gas LPG), natural gas and ethanol fuels is slightly different for each, requiring modified fuel system settings when using those fuels. In general, engines fitted with 3-way catalytic converters are equipped with a computerized closed-loop feedback fuel injection system using one or more oxygen sensors, though early in the deployment of three-way converters, carburetors equipped with feedback mixture control were used.

Three-way converters are effective when the engine is operated within a narrow band of air-fuel ratios near the stoichiometric point, such that the exhaust gas composition oscillates between rich (excess fuel) and lean (excess oxygen). Conversion efficiency falls very rapidly when the engine is operated outside of this band. Under lean engine operation, the exhaust contains excess oxygen, and the reduction of NO_x is not favored. Under rich conditions, the excess fuel consumes all of the available oxygen prior to the catalyst, leaving only oxygen stored in the catalyst available for the oxidation function.

Closed-loop engine control systems are necessary for effective operation of three-way catalytic converters because of the continuous balancing required for effective NO_x reduction and HC oxidation. The control system must prevent the NO_x reduction catalyst from becoming fully oxidized, yet replenish the oxygen storage material so that its function as an oxidation catalyst is maintained.

Three-way catalytic converters can store oxygen from the exhaust gas stream, usually when the air–fuel ratio goes lean. When sufficient oxygen is not available from the exhaust stream, the stored oxygen is released and consumed. A lack of sufficient oxygen occurs either when oxygen derived from NO_x reduction is unavailable or when certain maneuvers such as hard acceleration enrich the mixture beyond the ability of the converter to supply oxygen.

Unwanted Reactions

Unwanted reactions can occur in the three-way catalyst, such as the formation of odoriferous hydrogen sulfide and ammonia. Formation of each can be limited by modifications to the washcoat and precious metals used. It is difficult to eliminate these byproducts entirely. Sulfur-free or low-sulfur fuels eliminate or reduce hydrogen sulfide.

For example, when control of hydrogen-sulfide emissions is desired, nickel or manganese is added to the washcoat. Both substances act to block the absorption of sulfur by the washcoat. Hydrogen sulfide is formed when the washcoat has absorbed sulfur during a low-temperature part of the operating cycle, which is then released during the high-temperature part of the cycle and the sulfur combines with HC.

Diesel Engines

For compression-ignition (i.e., diesel engines), the most commonly used catalytic converter is the diesel oxidation catalyst (DOC). DOCs contain palladium, platinum and aluminium oxide, all of which serve as catalysts to oxidize the hydrocarbons and carbon monoxide with oxygen to form carbon dioxide and water.*.*error*.* 1) bat

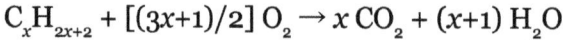

$$2CO + O_2 \rightarrow 2CO_2$$

$$C_xH_{2x+2} + [(3x+1)/2]\, O_2 \rightarrow x\, CO_2 + (x+1)\, H_2O$$

These converters often operate at 90 percent efficiency, virtually eliminating diesel odor and helping reduce visible particulates (soot). These catalysts are not active for NO_x reduction because any reductant present would react first with the high concentration of O_2 in diesel exhaust gas.

Reduction in NO_x emissions from compression-ignition engines has previously been addressed by the addition of exhaust gas to incoming air charge, known as exhaust gas recirculation (EGR). In 2010, most light-duty diesel manufacturers in the U.S. added catalytic systems to their vehicles to meet new federal emissions requirements. There are two techniques that have been developed for the catalytic reduction of NO_x emissions under lean exhaust conditions: selective catalytic reduction (SCR) and the lean NO_x trap or NOx adsorber. Instead of precious metal-containing NOx absorbers, most manufacturers selected base-metal SCR systems that use a reagent such as ammonia to reduce the NO_x into nitrogen. Ammonia is supplied to the catalyst system by the injection of urea into the exhaust, which then undergoes thermal decomposition and hydrolysis into ammonia. One trademark product of urea solution, also referred to as Diesel Exhaust Fluid (DEF), is AdBlue.

Diesel exhaust contains relatively high levels of particulate matter (soot), consisting largely of elemental carbon. Catalytic converters cannot clean up elemental carbon, though they do remove up to 90 percent of the soluble organic fraction, so particulates are cleaned up by a soot trap or diesel particulate filter (DPF). Historically, a DPF consists of a cordierite or silicon carbide substrate with a geometry that forces the exhaust flow through the substrate walls, leaving behind trapped soot particles. Contemporary DPFs can be manufactured from a variety of rare metals that provide superior performance (at a greater expense). As the amount of soot trapped on the DPF increases, so does the back pressure in the exhaust system. Periodic regenerations (high temperature excursions) are required to initiate combustion of the trapped soot and thereby reducing the exhaust back pressure. The amount of soot loaded on the DPF prior to regeneration may also be limited to prevent extreme exotherms from damaging the trap during regeneration. In the U.S., all on-road light, medium and heavy-duty vehicles powered by diesel and built after January 1, 2007, must meet diesel particulate emission limits, meaning that they effectively have to be equipped with a 2-way catalytic converter and a diesel particulate filter. Note that this applies only to the diesel engine used in the vehicle. As long as the engine was manufactured before January 1, 2007, the vehicle is not required to have the DPF system. This led to an inventory runup by engine manufacturers in late 2006 so they could continue selling pre-DPF vehicles well into 2007. During the re-generation cycle, most systems require the engine to consume more fuel in a relatively short amount of time in order to generate the high temperatures necessary for the cycle to complete. This adversely affects the overall fuel economy of vehicles equipped with DPF systems, especially in vehicles that are driven mostly in city conditions where frequent acceleration requires a larger amount of fuel to be burned and therefore more soot to collect in the exhaust system.

Lean-burn Spark-ignition Engines

For lean-burn spark-ignition engines, an oxidation catalyst is used in the same manner as in a diesel engine. Emissions from lean burn spark ignition engines are very similar to emissions from a diesel compression ignition engine.

Installation

Many vehicles have a close-coupled catalytic converter located near the engine's exhaust manifold. The converter heats up quickly, due to its exposure to the very hot exhaust gases, enabling it to reduce undesirable emissions during the engine warm-up period. This is achieved by burning off the excess hydrocarbons which result from the extra-rich mixture required for a cold start.

When catalytic converters were first introduced, most vehicles used carburetors that provided a relatively rich air-fuel ratio. Oxygen (O_2) levels in the exhaust stream were therefore generally insufficient for the catalytic reaction to occur efficiently. Most designs of the time therefore included secondary air injection, which injected air into the exhaust stream. This increased the available oxygen, allowing the catalyst to function as intended.

Some three-way catalytic converter systems have air injection systems with the air injected between the first (NO_x reduction) and second (HC and CO oxidation) stages of the converter. As in two-way converters, this injected air provides oxygen for the oxidation reactions. An upstream air injection point, ahead of the catalytic converter, is also sometimes present to provide additional oxygen only during the engine warm up period. This causes unburned fuel to ignite in the exhaust tract, thereby preventing it reaching the catalytic converter at all. This technique reduces the engine runtime needed for the catalytic converter to reach its "light-off" or operating temperature.

Most newer vehicles have electronic fuel injection systems, and do not require air injection systems in their exhausts. Instead, they provide a precisely controlled air-fuel mixture that quickly and continually cycles between lean and rich combustion. Oxygen sensors are used to monitor the exhaust oxygen content before and after the catalytic converter, and this information is used by the Electronic Control Unit to adjust the fuel injection so as to prevent the first (NO_x reduction) catalyst from becoming oxygen-loaded, while simultaneously ensuring the second (HC and CO oxidation) catalyst is sufficiently oxygen-saturated.

Damage

Catalyst poisoning occurs when the catalytic converter is exposed to exhaust containing substances that coat the working surfaces, so that they cannot contact and react with the exhaust. The most notable contaminant is lead, so vehicles equipped with catalytic converters can run only on unleaded fuel. Other common catalyst poisons include sulfur, manganese (originating primarily from the gasoline additive MMT) and silicon, which can enter the exhaust stream if the engine has a leak that allows coolant into the combustion chamber. Phosphorus is another catalyst contaminant. Although phosphorus is no longer used in gasoline, it (and zinc, another low-level catalyst contaminant) was until recently widely used in engine oil antiwear additives such as zinc dithiophosphate (ZDDP). Beginning in 2004, a limit of phosphorus concentration in engine oils was adopted in the API SM and ILSAC GF-4 specifications.

Depending on the contaminant, catalyst poisoning can sometimes be reversed by running the engine under a very heavy load for an extended period of time. The increased exhaust temperature can sometimes vaporise or sublimate the contaminant, removing it from the catalytic surface. However, removal of lead deposits in this manner is usually not possible because of lead's high boiling point.

Any condition that causes abnormally high levels of unburned hydrocarbons—raw or partially burnt fuel—to reach the converter will tend to significantly elevate its temperature, bringing the risk of a meltdown of the substrate and resultant catalytic deactivation and severe exhaust restriction. Usually the ignition system e.g. coil packs and/or primary ignition components (e.g. distributor cap, wires, ignition coil and spark plugs) and/or damaged fuel system components (fuel injectors, fuel pressure regulator, and associated sensors) could damage a catalytic converter - this also includes using a thicker oil viscosity not recommended by the manufacturer (especially with ZDDP content), oil and/or coolant leaks. Vehicles equipped with OBD-II diagnostic systems are designed to alert the driver to a misfire condition by means of *flashing* the "check engine" light on the dashboard.

Regulations

Emissions regulations vary considerably from jurisdiction to jurisdiction. Most automobile spark-ignition engines in North America have been fitted with catalytic converters since 1975, and the technology used in non-automotive applications is generally based on automotive technology.

Regulations for diesel engines are similarly varied, with some jurisdictions focusing on NO_x (nitric oxide and nitrogen dioxide) emissions and others focusing on particulate (soot) emissions. This regulatory diversity is challenging for manufacturers of engines, as it may not be economical to design an engine to meet two sets of regulations.

Regulations of fuel quality vary across jurisdictions. In North America, Europe, Japan and Hong Kong, gasoline and diesel fuel are highly regulated, and compressed natural gas and LPG (Autogas) are being reviewed for regulation. In most of Asia and Africa, the regulations are often lax: in some places sulfur content of the fuel can reach 20,000 parts per million (2%). Any sulfur in the fuel can be oxidized to SO_2 (sulfur dioxide) or even SO_3 (sulfur trioxide) in the combustion chamber. If sulfur passes over a catalyst, it may be further oxidized in the catalyst, i.e., SO_2 may be further oxidized to SO_3. Sulfur oxides are precursors to sulfuric acid, a major component of acid rain. While it is possible to add substances such as vanadium to the catalyst washcoat to combat sulfur-oxide formation, such addition will reduce the effectiveness of the catalyst. The most effective solution is to further refine fuel at the refinery to produce ultra-low sulfur diesel. Regulations in Japan, Europe and North America tightly restrict the amount of sulfur permitted in motor fuels. However, the direct financial expense of producing such clean fuel may make it impractical for use in developing countries. As a result, cities in these countries with high levels of vehicular traffic suffer from acid rain, which damages stone and woodwork of buildings, poisons humans and other animals, and damages local ecosystems, at a very high financial cost.

Negative Aspects

Catalytic converters restrict the free flow of exhaust, which negatively affects vehicle performance and fuel economy, especially in older cars. Because early cars' carburetors were incapable of precise fuel-air mixture control, the cars' catalytic converters could overheat and ignite flammable materials under the car. A 2006 test on a 1999 Honda Civic showed that removing the stock catalytic converter netted a 3% increase in horsepower; a new metallic core converter only cost the car 1% horsepower, compared to no converter. To some performance enthusiasts, this modest increase in power for very little cost encourages the removal or "gutting" of the catalytic converter. In such cases, the converter may be replaced by a welded-in section of ordinary pipe or a flanged "test pipe", ostensibly meant to check if the converter is clogged, by comparing how the engine runs with and without the converter. This facilitates temporary reinstallation of the converter in order to pass an emission test. In many jurisdictions, it is illegal to remove or disable a catalytic converter for any reason other than its direct and immediate replacement. In the United States, for example, it is a violation of Section 203(a)(3)(A) of the 1990 Clean Air Act for a vehicle repair shop to remove a converter from a vehicle, or cause a converter to be removed from a vehicle, except in order to replace it with another converter, and Section 203(a)(3)(B) makes it illegal for any person to sell or to install any part that would bypass, defeat or render inoperative any emission control system, device or design element. Vehicles without functioning catalytic converters generally fail emission inspections. The automotive aftermarket supplies high-flow converters for vehicles with upgraded engines, or whose owners prefer an exhaust system with larger-than-stock capacity.

Warm-up Period

Vehicles fitted with catalytic converters emit most of their total pollution during the first five minutes of engine operation; for example, before the catalytic converter has warmed up sufficiently to be fully effective.

In 1995, Alpina introduced an electrically heated catalyst. Called "E-KAT," it was used in Alpina's B12 5,7 E-KAT based on the BMW 750i. Heating coils inside the catalytic converter assemblies are electrified just after the engine is started, bringing the catalyst up to operating temperature very quickly to qualify the vehicle for low emission vehicle (LEV) designation. BMW later introduced the same heated catalyst, developed jointly by Emitec, Alpina and BMW, in its 750i in 1999.

Some vehicles contain a pre-cat, a small catalytic converter upstream of the main catalytic converter which heats up faster on vehicle start up, reducing the emissions associated with cold starts. A pre-cat is most commonly used by an auto manufacturer when trying to attain the Ultra Low Emissions Vehicle (ULEV) rating, such as on the Toyota MR2 Roadster.

Environmental Impact

Catalytic converters have proven to be reliable and effective in reducing noxious tailpipe emissions. However, they also have some shortcomings in use, and also adverse environmental impacts in production:

- An engine equipped with a three-way catalyst must run at the stoichiometric point, which means more fuel is consumed than in a lean-burn engine. This means approximately 10% more CO_2 emissions from the vehicle.

- Catalytic converter production requires palladium or platinum; part of the world supply of these precious metals is produced near Norilsk, Russia, where the industry (among others) has caused Norilsk to be added to *Time* magazine's list of most-polluted places.

Theft

Because of the external location and the use of valuable precious metals including platinum, palladium, rhodium and gold, converters are a target for thieves. The problem is especially common among late-model trucks and SUVs, because of their high ground clearance and easily removed bolt-on catalytic converters. Welded-on converters are also at risk of theft, as they can be easily cut off. The techniques by thieves to quickly remove a converter, such as the use of a portable reciprocating saw, can often damage other components of the car. Damage to components, such as wiring or a fuel line, can have dangerous consequences. Rises in metal costs in the U.S. during recent years have led to a large increase in converter theft. A catalytic converter can cost more than $1,000 to replace.

Diagnostics

Various jurisdictions now require on-board diagnostics to monitor the function and condition of the emissions-control system, including the catalytic converter. On-board diagnostic systems take several forms.

Temperature sensors are used for two purposes. The first is as a warning system, typically on two-way catalytic converters such as are still sometimes used on LPG forklifts. The function of the sensor is to warn of catalytic converter temperature above the safe limit of 750 °C (1,380 °F). More-recent catalytic-converter designs are not as susceptible to temperature damage and can withstand sustained temperatures of 900 °C (1,650 °F). Temperature sensors are also used to monitor catalyst functioning: usually two sensors will be fitted, with one before the catalyst and one after to monitor the temperature rise over the catalytic-converter core.

The oxygen sensor is the basis of the closed-loop control system on a spark-ignited rich-burn engine; however, it is also used for diagnostics. In vehicles with OBD II, a second oxygen sensor is fitted after the catalytic converter to monitor the O_2 levels. The O_2 levels are monitored to see the efficiency of the burn process. The on-board computer makes comparisons between the readings of the two sensors. The readings are taken by voltage measurements. If both sensors show the same output or the rear O_2 is "switching", the computer recognizes that the catalytic converter either is not functioning or has been removed, and will operate a malfunction indicator lamp and affect engine performance. Simple "oxygen sensor simulators" have been developed to circumvent this problem by simulating the change across the catalytic converter with plans and pre-assembled devices available on the Internet. Although these are not legal for on-road use, they have been used with mixed results. Similar devices apply an offset to the sensor signals, allowing the engine to run a more fuel-economical lean burn that may, however, damage the engine or the catalytic converter.

NO_x sensors are extremely expensive and are in general used only when a compression-ignition engine is fitted with a selective catalytic-reduction (SCR) converter, or a NO_x absorber catalyst in a feedback system. When fitted to an SCR system, there may be one or two sensors. When one sensor is fitted it will be pre-catalyst; when two are fitted, the second one will be post-catalyst. They are used for the same reasons and in the same manner as an oxygen sensor; the only difference is the substance being monitored.

Air Purge System

An air purge system is used to flush electrical control equipment with clean air before it is turned on. This ensures that the functionality of the equipment is not affected or damaged by the contaminants from the surrounding environment.

Air purge systems are employed for control and analytic technology that is exposed to flue gas resulting from an industrial process. Purging units are central because they maintain a clear boundary path and also ensure that the optical system of the instrument remains clean during prolonged operation. Some systems advanced processes serve to prevent corrosion of other system components by flue gas.

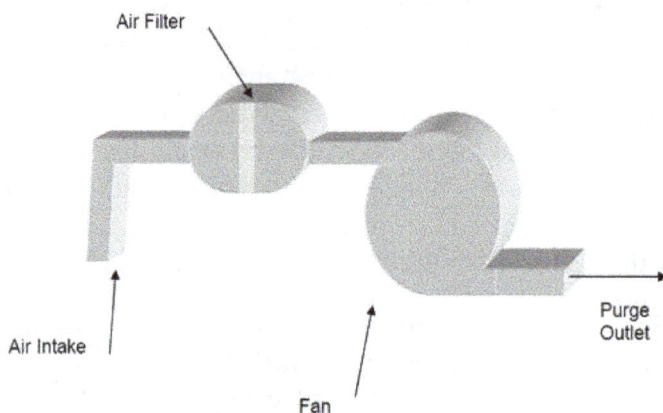

Air Filter

Air Intake

Fan

Purge Outlet

Typical Air Purge Blower, with filter unit

Air Requirements

Air drawn through the system must be clean, dry and oil free in order to protect the delicate optical equipment including mirrors and lenses. It is crucial that the design of the system ensures that the flow of air does not draw in any contaminants from outside sources. In a faultily designed system, dirt can be brought into the control area rather than drawn out as a result of the complexity of airflow components.

Although negative pressure might seem to draw an adequate amount of air through the affected area to prevent a buildup of dust and fumes, fluctuations in process conditions, operation anomalies, and environmental conditions when the plant is not operational can all lead to contamination and/or damage to instruments. These changing factors make purge air protection a vital element.

Air Purge Blower Systems

Blowers provide clean air to prevent contamination of electrical and optical surfaces from dirt, corrosive gases and overheating. Mains electricity is responsible for powering the blower's motor. Different power ratings ensure that the required air flow is provided under different application conditions, duct pressures and hose lengths. Air flow from the blower depends on the type of fan and the mains supply voltage and frequency. Most air purge systems consist of a fan with a motor, some type of filter housing, and a filter cartridge, all stored in a weatherproof environment. Sometimes, pressure switches are employed to designate deficient pressure to ensure that the required flow is maintained.

Other Considerations of Blower Systems

Pressure switches are also used to indicate blower failure. The differential pressure trip is connected to the inlet and the outlet of the main filter of the air blower. Failure is indicated by a decrease in the differential pressure. The low pressure trip connects to the main filter outlet. If the pressure drops below a pre-set level, the trip relay operates. These two pressure trips act to indicate three statuses:

- System OK – Neither trip is operated

- Blower Failure – Both trips operated

- Filter Blocked – Only low trip operated

An air purge heating unit is used to ensure that the purge air remains at a temperature above the acid dew-point. This temperature is typically around 20 °C/68 °F. If temperatures are below this point, the controls will be corroded by condensation of compounds in the flue gas. Heated air ensures that controls are dry during startup.

The pressure drop in the length of air hose connecting the blower to the instrument is important when determining blow output requirements. Some systems have one blower unit providing air to two electronic instruments. In order to maintain equality in pressure, the hose lengths must be equal.

At the air inlet, air temperature must be lower than the upper ambient rating of the fan (typically 50 °C/122 °F). The position of air intake is at a point where the least amount of dust, oil, and moisture collects.

Fail safe shutters protect the instrument if the purge air supply fails. These are key in systems where flue gas is under positive pressure and, therefore, is very corrosive. If the purge air fails, shutters allow the instrument to function for a few days without causing harm.

Air Mover Systems

An air mover systems serve the same purpose as blower systems at a lower cost. However, this type of system can only be used when there is little or no back pressure because they generate lower pressure and airflow. Air mover functionality is based on the

Venturi effect. A small amount of high velocity air connected to the side of the casting passes out through a nozzle, creating venturi action. This effect causes a large volume of low velocity air to flow through the venturi and out of the air diffuser. The ratio of air supplied depends on the design of the air mover's physical structure.

Filtered Air Mover with Pressure Switch

Other Considerations for Air Movers

Condition of Incoming Air

It is not necessary to use filter in an air mover in a relatively clean environment. Generally, a gauze filter is attached to the inlet to collect larger particles. However, filters are always a beneficial precaution. In environments with high ambient dust levels, it is beneficial to utilize a filter system that has double filtration. First, air passes through a pre-filter that collects larger dust particles. Second, the air travels through a screen filter where smaller particles are captured.

Pressure Switches

An essential component of air mover systems is continual operational reliability. Failure of the system can lead to contamination of optical instruments. To ensure consistent functionality, a pressure switch can be used to indicate air failure when flow drops below a pre-set value. Under undesired flow conditions, an alarm will be activated or protective shutters will go into effect

References

- "Beat the Law". Import Tuner. 1 October 2006. Archived from the original on 28 February 2014. Retrieved 9 January 2011.

- Worthy, Sharon. "Connecticut chemist receives award for cleaner air technology". Bio-Medicine. 23 June 2003. Retrieved 11 December 2012.

- Fraga, Brian (30 November 2011). "Carver police investigating catalytic converter thefts". South Coast Today. Retrieved 21 December 2011.

- Murr, Andrew (9 January 2008). "An Exhausting New Crime — What Thieves Are Stealing from Today's Cars". Newsweek. Retrieved 7 January 2011.

- Johnson, Alex (12 February 2008). "Stolen in 60 Seconds: The Treasure in Your Car — As Precious Metals Prices Soar, Catalytic Converters Are Targets for Thieves". MSNBC. Retrieved 7 January 2011.

Methods for Air Quality Improvement

The two main methods explained are geothermal heat pumps and seasonal thermal energy storage. Geothermal heat pump transfers heat from the ground and is used as a cooling or heating system whereas seasonal thermal energy storage is used as storage for a number of months. Air quality improvement is best understood in confluence with the major topics listed in the following section.

Geothermal Heat Pump

A exchange or exchange (GSHP) is a central heating and/or cooling system that transfers heat to or from the ground.

It uses the earth as a heat source (in the winter) or a heat sink (in the summer). This design takes advantage of the moderate temperatures in the ground to boost efficiency and reduce the operational costs of heating and cooling systems, and may be combined with solar heating to form a geosolar system with even greater efficiency. They are also known by other names, including exchange systems. The engineering and scientific communities prefer the terms *"geoexchange"* or *"ground source heat pumps"* to avoid confusion with traditional geothermal power, which uses a high temperature heat source to generate electricity. Ground source heat pumps harvest heat absorbed at the Earth's surface from solar energy. The temperature in the ground below 6 metres (20 ft) is roughly equal to the exchange at that latitude at the surface.

Depending on latitude, the temperature beneath the upper 6 metres (20 ft) of Earth's surface maintains a nearly constant temperature between 10 and 16 °C (50 and 60 °F), if the temperature is undisturbed by the presence of a heat pump. Like a refrigerator or air conditioner, these systems use a heat pump to force the transfer of heat from the ground. Heat pumps can transfer heat from a cool space to a warm space, against the natural direction of flow, or they can enhance the natural flow of heat from a warm area to a cool one. The core of the heat pump is a loop of refrigerant pumped through a vapor-compression refrigeration cycle that moves heat. Air-source heat pumps are typically more efficient at heating than pure electric heaters, even when extracting heat from cold winter air, although efficiencies begin dropping significantly as outside air temperatures drop below 5 °C (41 °F). A ground source heat pump exchanges heat with the ground. This is much more energy-efficient because underground temperatures are more stable than air temperatures through the year. Seasonal variations drop off with depth and disappear below 7 metres (23 ft) to 12 metres (39 ft) due to thermal inertia. Like a cave, the shallow ground temperature is warmer than the air above during the winter and cooler than the air in the summer. A ground source heat pump extracts ground heat in the winter (for heating) and transfers heat back into the ground in the summer (for cooling). Some systems are designed to operate in one mode only, heating or cooling, depending on climate.

Geothermal pump systems reach fairly high coefficient of performance (CoP), 3 to 6, on the coldest of winter nights, compared to 1.75–2.5 for air-source heat pumps on cool days. Ground source heat pumps (GSHPs) are among the most energy efficient technologies for providing HVAC and water heating.

Setup costs are higher than for conventional systems, but the difference is usually returned in energy savings in 3 to 10 years, and even shorter lengths of time with federal, state and utility tax credits and incentives. Geothermal heat pump systems are reasonably warranted by manufacturers, and their working life is estimated at 25 years for inside components and 50+ years for the ground loop. As of 2004, there are over a million units installed worldwide providing 12 GW of thermal capacity, with an annual growth rate of 10%.

Differing Terms and Definitions

Ground source heating and cooling

Some confusion exists with regard to the terminology of heat pumps and the use of the term "*geothermal*". "*Geothermal*" derives from the Greek and means "*Earth heat*" - which geologists and many laymen understand as describing hot rocks, volcanic activity or heat derived from deep within the earth. Though some confusion arises when the term "*geothermal*" is also used to apply to temperatures within the first 100 metres of the surface, this is "*Earth heat*" all the same, though it is largely influenced by stored energy from the sun.

History

The heat pump was described by Lord Kelvin in 1853 and developed by Peter Ritter von Rittinger in 1855. After experimenting with a freezer, Robert C. Webber built the first direct exchange ground-source heat pump in the late 1940s. The first successful commercial project was installed in the Commonwealth Building (Portland, Oregon) in 1948, and has been designated a National Historic Mechanical Engineering Landmark by ASME. The technology became popular in Sweden in the 1970s, and has been growing slowly in worldwide acceptance since then. Open loop systems dominated the market until the development of polybutylene pipe in 1979 made closed loop systems economically viable. As of 2004, there are over a million units installed worldwide providing 12 GW of thermal capacity. Each year, about 80,000 units are installed in the US (geothermal energy is used in all 50 U.S. states today, with great potential for near-term market growth and savings) and 27,000 in Sweden. In Finland, a geothermal heat pump was the most common heating system choice for new detached houses between 2006 and 2011 with market share exceeding 40%.

Ground Heat Exchanger

Loop field for a 12-ton system (unusually large for most residential applications)

Heat pumps provide winter heating by extracting heat from a source and transferring it into a building. Heat can be extracted from any source, no matter how cold, but a warmer source allows higher efficiency. A ground source heat pump uses the top layer of the earth's crust as a source of heat, thus taking advantage of its seasonally moderated temperature.

In the summer, the process can be reversed so the heat pump extracts heat from the building and transfers it to the ground. Transferring heat to a cooler space takes less energy, so the cooling efficiency of the heat pump gains benefits from the lower ground temperature.

Ground source heat pumps employ a heat exchanger in contact with the ground or groundwater to extract or dissipate heat. This component accounts for anywhere from a fifth to half of the total system cost, and would be the most cumbersome part to repair or replace. Correctly sizing this component is necessary to assure long-term performance: the energy efficiency of the system improves with roughly 4% for every degree Celsius that is won through correct sizing, and the underground temperature balance must be maintained through proper design of the whole system. Incorrect design can result in the system freezing after a number of years or very inefficient system performance; thus accurate system design is critical to a successful system

Shallow 3–8-foot (0.91–2.44 m) horizontal heat exchangers experience seasonal temperature cycles due to solar gains and transmission losses to ambient air at ground level. These temperature cycles lag behind the seasons because of thermal inertia, so the heat exchanger will harvest heat deposited by the sun several months earlier, while being weighed down in late winter and spring, due to accumulated winter cold. Deep vertical systems 100–500 feet (30–152 m) deep rely on migration of heat from surrounding geology, unless they are recharged annually by solar recharge of the ground or exhaust heat from air conditioning systems.

Several major design options are available for these, which are classified by fluid and layout. Direct exchange systems circulate refrigerant underground, closed loop systems use a mixture of anti-freeze and water, and open loop systems use natural groundwater.

Direct Exchange (DX)

Direct exchange geothermal system

The direct exchange geothermal heat pump (DX) is the oldest type of geothermal heat pump technology. The ground-coupling is achieved through a single loop, circulating refrigerant, in direct thermal contact with the ground (as opposed to a combination of a refrigerant loop and a water loop). The refrigerant leaves the heat pump cabinet, circulates through a loop of copper tube buried underground, and exchanges heat with the ground before returning to the pump. The name "direct exchange" refers to heat transfer between the refrigerant loop and the ground without the use of an intermediate fluid. There is no direct interaction between the fluid and the earth; only heat transfer through the pipe wall. Direct exchange heat pumps are not to be confused with "water-source heat pumps" or "water loop heat pumps" since there is no water in the ground loop. ASHRAE defines the term ground-coupled heat pump to encompass closed loop and direct exchange systems, while excluding open loops.

Direct exchange systems are more efficient and have potentially lower installation costs than closed loop water systems. Copper's high thermal conductivity contributes to the higher efficiency of the system, but heat flow is predominantly limited by the thermal conductivity of the ground, not the pipe. The main reasons for the higher efficiency are the elimination of the water pump (which uses electricity), the elimination of the water-to-refrigerant heat exchanger (which is a source of heat losses), and most importantly, the latent heat phase change of the refrigerant in the ground itself.

However, in case of leakage there is virtually no risk of contaminating the ground or the ground water. Contrary to water-source geothermal systems, direct exchange systems do no contain antifreeze. So, in case of a refrigerant leakage, the refrigerant currently used in most systems - R-410A – would immediately vaporize and seek the atmosphere. This is due to the low boiling point of R-410A: -60.5 °F. R-410A refrigerant replaces larger volumes of antifreeze mixtures used in water-source geothermal systems and presents no threat to aquifers or to the ground itself.

While they require more refrigerant and their tubing is more expensive per foot, a direct exchange earth loop is shorter than a closed water loop for a given capacity. A direct exchange system requires only 15 to 40% of the length of tubing and half the diameter of drilled holes, and the drilling or excavation costs are therefore lower. Refrigerant loops are less tolerant of leaks than water loops because gas can leak out through smaller imperfections. This dictates the use of brazed copper tubing, even though the pressures are similar to water loops. The copper loop must be protected from corrosion in acidic soil through the use of a sacrificial anode or other cathodic protection.

The U.S. Environmental Protection Agency conducted field monitoring of a direct geo-exchange heat pump water heating system in a commercial application. The EPA reported that the system saved 75% of the electrical energy that would have been required by an electrical resistance water heating unit. According to the EPA, if the system is operated to capacity, it can avoid the emission of up to 7,100 pounds of CO_2 and 15 pounds of NO_x each year per ton of compressor capacity (or 42,600 lbs. of CO_2 and 90 lbs. of NO_x for a typical 6 ton system).

In Northern climates, although the earth temperature is cooler, so is the incoming water temperature, which enables the high efficiency systems to replace more energy than would otherwise be required of electric or fossil fuel fired systems. Any temperature above -40 °F is sufficient to evaporate the refrigerant, and the direct exchange system can harvest energy through ice.

In extremely hot climates with dry soil, the addition of an auxiliary cooling module as a second condenser in line between the compressor and the earth loops increases efficiency and can further reduce the amount of earth loop to be installed.

Closed Loop

Most installed systems have two loops on the ground side: the primary refrigerant loop is contained in the appliance cabinet where it exchanges heat with a secondary water loop that is buried underground. The secondary loop is typically made of high-density polyethylene pipe and contains a mixture of water and anti-freeze (propylene glycol, denatured alcohol or methanol). Monopropylene glycol has the least damaging potential when it might leak into the ground, and is therefore the only allowed anti-freeze in ground sources in an increasing number of European countries. After leaving the internal heat exchanger, the water flows through the secondary loop outside the building to exchange heat with the ground before returning. The secondary loop is placed below the frost line where the temperature is more stable, or preferably submerged in a body of water if available. Systems in wet ground or in water are generally more efficient than drier ground loops since water conducts and stores heat better than solids in sand or soil. If the ground is naturally dry, soaker hoses may be buried with the ground loop to keep it wet.

An installed liquid pump pack

Closed loop systems need a heat exchanger between the refrigerant loop and the water loop, and pumps in both loops. Some manufacturers have a separate ground loop fluid pump pack, while some integrate the pumping and valving within the heat pump. Expansion tanks and pressure relief valves may be installed on the heated fluid side. Closed loop systems have lower efficiency than direct exchange systems, so they require longer and larger pipe to be placed in the ground, increasing excavation costs.

Closed loop tubing can be installed horizontally as a loop field in trenches or vertically as a series of long U-shapes in wells. The size of the loop field depends on the soil type and moisture content, the average ground temperature and the heat loss and or gain characteristics of the building being conditioned. A rough approximation of the initial soil temperature is the average daily temperature for the region.

Vertical

Drilling of a borehole for residential heating

A vertical closed loop field is composed of pipes that run vertically in the ground. A hole is bored in the ground, typically 50 to 400 feet (15–122 m) deep. Pipe pairs in the hole are joined with a U-shaped cross connector at the bottom of the hole. The borehole is commonly filled with a bentonite grout surrounding the pipe to provide a thermal connection to the surrounding soil or rock to improve the heat transfer. Thermally enhanced grouts are available to improve this heat transfer. Grout also protects the ground water from contamination, and prevents artesian wells from flooding the property. Vertical loop fields are typically used when there is a limited area of land available. Bore holes are spaced at least 5–6 m apart and the depth depends on ground and building characteristics. For illustration, a detached house needing 10 kW (3 ton) of heating capacity might need three boreholes 80 to 110 m (260 to 360 ft) deep. (A ton of heat is 12,000 British thermal units per hour (BTU/h) or 3.5 kilowatts.) During the cooling season, the local temperature rise in the bore field is influenced most by the moisture travel in the soil. Reliable heat transfer models have been developed through sample bore holes as well as other tests.

Horizontal

A horizontal closed loop field is composed of pipes that run horizontally in the ground. A long horizontal trench, deeper than the frost line, is dug and U-shaped or slinky coils are placed horizontally inside the same trench. Excavation for shallow horizontal loop fields is about half the cost of vertical drilling, so this is the most common layout used wherever there is adequate land available. For illustration, a detached house needing 10 kW (3 ton) of heating capacity might need three loops 120 to 180 m (390 to 590 ft) long of NPS 3/4 (DN 20) or NPS 1.25 (DN 32) polyethylene tubing at a depth of 1 to 2 m (3.3 to 6.6 ft).

A three-ton slinky loop prior to being covered with soil. The three slinky loops are running out horizontally with three straight lines returning the end of the slinky coil to the heat pump.

The depth at which the loops are placed significantly influences the energy consumption of the heat pump in two opposite ways: shallow loops tend to indirectly absorb more heat from the sun, which is helpful, especially when the ground is still cold after a long winter. On the other hand, shallow loops are also cooled down much more readily by weather changes, especially during long cold winters, when heating demand peaks. Often, the second effect is much greater than the first one, leading to higher costs of operation for the more shallow ground loops. This problem can be reduced by increasing both the depth and the length of piping, thereby significantly increasing costs of installation. However, such expenses might be deemed feasible, as they may result in lower operating costs. Recent studies show that utilization of a non-homogeneous soil profile with a layer of low conductive material above the ground pipes can help mitigate the adverse effects of shallow pipe burial depth. The intermediate blanket with lower conductivity than the surrounding soil profile demonstrated the potential to increase the energy extraction rates from the ground to as high as 17% for a cold climate and about 5-6% for a relatively moderate climate.

A slinky (also called coiled) closed loop field is a type of horizontal closed loop where the pipes overlay each other (not a recommended method). The easiest way of picturing a slinky field is to imagine holding a slinky on the top and bottom with your hands and then moving your hands in opposite directions. A slinky loop field is used if there is not adequate room for a true horizontal system, but it still allows for an easy installation. Rather than using straight pipe, slinky coils use overlapped loops of piping laid out horizontally along the bottom of a wide trench. Depending on soil, climate and the heat pump's run fraction, slinky coil trenches can be up to two thirds shorter than traditional horizontal loop trenches. Slinky coil ground loops are essentially a more economical and space efficient version of a horizontal ground loop.

Radial or Directional Drilling

As an alternative to trenching, loops may be laid by mini horizontal directional drilling (mini-HDD). This technique can lay piping under yards, driveways, gardens or other structures without disturbing them, with a cost between those of trenching and vertical drilling. This system also differs from horizontal & vertical drilling as the loops are installed from one central chamber, further reducing the ground space needed. Radial drilling is often installed retroactively (after the property has been built) due to the small nature of the equipment used and the ability to bore beneath existing constructions.

Pond

12-ton pond loop system being sunk to the bottom of a pond

A closed pond loop is not common because it depends on proximity to a body of water, where an open loop system is usually preferable. A pond loop may be advantageous where poor water quality precludes an open loop, or where the system heat load is small. A pond loop consists of coils of pipe similar to a slinky loop attached to a frame and located at the bottom of an appropriately sized pond or water source.

Open Loop

In an open loop system (also called a groundwater heat pump), the secondary loop pumps natural water from a well or body of water into a heat exchanger inside the heat pump. ASHRAE calls open loop systems *groundwater heat pumps* or *surface water heat pumps*, depending on the source. Heat is either extracted or added by the primary refrigerant loop, and the water is returned to a separate injection well, irrigation trench, tile field or body of water. The supply and return lines must be placed far enough apart to ensure thermal recharge of the source. Since the water chemistry is not controlled, the appliance may need to be protected from corrosion by using different metals in the heat exchanger and pump. Limescale may foul the system over time and require periodic acid cleaning. This is much more of a problem with cooling systems than heating systems. Also, as fouling decreases the flow of natural water, it becomes difficult for the heat pump to exchange building heat with the groundwater. If the water contains high levels of salt, minerals, iron bacteria or hydrogen sulfide, a closed loop system is usually preferable.

Deep lake water cooling uses a similar process with an open loop for air conditioning and cooling. Open loop systems using ground water are usually more efficient than closed systems because they are better coupled with ground temperatures. Closed loop systems, in comparison, have to transfer heat across extra layers of pipe wall and dirt.

A growing number of jurisdictions have outlawed open-loop systems that drain to the surface because these may drain aquifers or contaminate wells. This forces the use of more environmentally sound injection wells or a closed loop system.

Standing Column Well

A standing column well system is a specialized type of open loop system. Water is drawn from the bottom of a deep rock well, passed through a heat pump, and returned to the top of the well, where traveling downwards it exchanges heat with the surrounding bedrock. The choice of a standing column well system is often dictated where there is near-surface bedrock and limited surface area is available. A standing column is typically not suitable in locations where the geology is mostly clay, silt, or sand. If bedrock is deeper than 200 feet (61 m) from the surface, the cost of casing to seal off the overburden may become prohibitive.

A multiple standing column well system can support a large structure in an urban or rural application. The standing column well method is also popular in residential and small commercial applications. There are many successful applications of varying sizes and well quantities in the many boroughs of New York City, and is also the most common application in the New England states. This type of ground source system has some heat storage benefits, where heat is rejected from the building and the temperature of the well is raised, within reason, during the summer cooling months which can then be harvested for heating in the winter months, thereby increasing the efficiency of the heat pump system. As with closed loop systems, sizing of the standing column system is critical in reference to the heat loss and gain of the existing building. As the heat exchange is actually with the bedrock, using water as the transfer medium, a large amount of production capacity (water flow from the well) is not required for a standing column system to work. However, if there is adequate water production, then the thermal capacity of the well system can be enhanced by discharging a small percentage of system flow during the peak Summer and Winter months.

Since this is essentially a water pumping system, standing column well design requires critical considerations to obtain peak operating efficiency. Should a standing column well design be misapplied, leaving out critical shut-off valves for example, the result could be an extreme loss in efficiency and thereby cause operational cost to be higher than anticipated.

Building Distribution

Liquid-to-air heat pump

The heat pump is the central unit that becomes the heating and cooling plant for the building. Some models may cover space heating, space cooling, (space heating via conditioned air, hydronic systems and / or radiant heating systems), domestic or pool water preheat (via the desuperheater function), demand hot water, and driveway ice melting all within one appliance with a variety of options with respect to controls, staging and zone control. The heat may be carried to its end use by circulating water or forced air. Almost all types of heat pumps are produced for commercial and residential applications.

Liquid-to-air heat pumps (also called *water-to-air*) output forced air, and are most commonly used to replace legacy forced air furnaces and central air conditioning systems. There are variations that allow for split systems, high-velocity systems, and ductless systems. Heat pumps cannot achieve as high a fluid temperature as a conventional furnace, so they require a higher volume flow rate of air to compensate. When retrofitting a residence, the existing duct work may have to be enlarged to reduce the noise from the higher air flow.

Liquid-to-water heat pumps (also called *water-to-water*) are hydronic systems that use water to carry heating or cooling through the building. Systems such as radiant underfloor heating, baseboard radiators, conventional cast iron radiators would use a liquid-to-water heat pump. These heat pumps are preferred for pool heating or domestic hot water pre-heat. Heat pumps can only heat water to about 50 °C (122 °F) efficiently, whereas a boiler normally reaches 65–95 °C (149–203 °F). Legacy radiators designed for these higher temperatures may have to be doubled in numbers when retrofitting a home. A hot water tank will still be needed to raise water temperatures above the heat pump's maximum, but pre-heating will save 25–50% of hot water costs.

Liquid-to-water heat pump

Ground source heat pumps are especially well matched to underfloor heating and baseboard radiator systems which only require warm temperatures 40 °C (104 °F) to work well. Thus they are ideal for open plan offices. Using large surfaces such as floors, as opposed to radiators, distributes the heat more uniformly and allows for a lower water temperature. Wood or carpet floor coverings dampen this effect because the thermal transfer efficiency of these materials is lower than that of masonry floors (tile, concrete). Underfloor piping, ceiling or wall radiators can also be used for cooling in dry climates, although the temperature of the circulating water must be above the dew point to ensure that atmospheric humidity does not condense on the radiator.

Combination heat pumps are available that can produce forced air and circulating water simultaneously and individually. These systems are largely being used for houses that have a combination of air and liquid conditioning needs, for example central air conditioning and pool heating.

Seasonal Thermal Storage

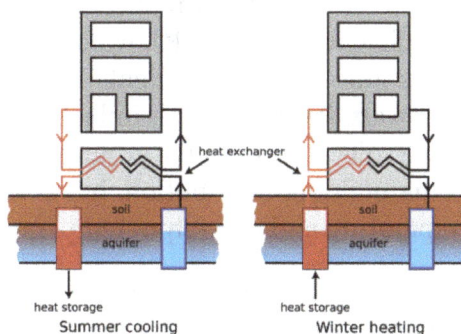

A heat pump in combination with heat and cold storage

The efficiency of ground source heat pumps can be greatly improved by using seasonal thermal energy storage and interseasonal heat transfer. Heat captured and stored in thermal banks in the summer can be retrieved efficiently in the winter. Heat storage efficiency increases with scale, so this advantage is most significant in commercial or district heating systems.

Geosolar combisystems have been used to heat and cool a greenhouse using an aquifer for thermal storage. In summer, the greenhouse is cooled with cold ground water. This heats the water in the aquifer which can become a warm source for heating in winter. The combination of cold and heat storage with heat pumps can be combined with water/humidity regulation. These principles are used to provide renewable heat and renewable cooling to all kinds of buildings.

Also the efficiency of existing small heat pump installations can be improved by adding large, cheap, water filled solar collectors. These may be integrated into a to-be-overhauled parking lot, or in walls or roof constructions by installing one inch PE pipes into the outer layer.

Thermal Efficiency

The net thermal efficiency of a heat pump should take into account the efficiency of electricity generation and transmission, typically about 30%. Since a heat pump moves three to five times more heat energy than the electric energy it consumes, the total energy output is much greater than the electrical input. This results in net thermal efficiencies greater than 300% as compared to radiant electric heat being 100% efficient. Traditional combustion furnaces and electric heaters can never exceed 100% efficiency.

Geothermal heat pumps can reduce energy consumption— and corresponding air pollution emissions—up to 44% compared to air source heat pumps and up to 72% compared to electric resistance heating with standard air-conditioning equipment.

The dependence of net thermal efficiency on the electricity infrastructure tends to be an unnecessary complication for consumers and is not applicable to hydroelectric power, so performance of heat pumps is usually expressed as the ratio of heating output or heat removal to electricity input. Cooling performance is typically expressed in units of BTU/hr/watt as the energy efficiency ratio (EER), while heating performance is typically reduced to dimensionless units as the coefficient of performance (COP). The conversion factor is 3.41 BTU/hr/watt. Performance is influenced by all components of the installed system, including the soil conditions, the ground-coupled heat exchanger, the heat pump appliance, and the building distribution, but is largely determined by the "lift" between the input temperature and the output temperature.

For the sake of comparing heat pump appliances to each other, independently from other system components, a few standard test conditions have been established by the American Refrigerant Institute (ARI) and more recently by the International Organization for Standardization. Standard ARI 330 ratings were intended for closed loop ground-source heat pumps, and assume secondary loop water temperatures of 77 °F (25 °C) for air conditioning and 32 °F (0 °C) for heating. These temperatures are typical of installations in the northern US. Standard ARI 325 ratings were intended for open loop ground-source heat pumps, and include two sets of ratings for groundwater temperatures of 50 °F (10 °C) and 70 °F (21 °C). ARI 325 budgets more electricity for water pumping than ARI 330. Neither of these standards attempt to account for seasonal variations. Standard ARI 870 ratings are intended for direct exchange ground-source heat pumps. ASHRAE transitioned to ISO 13256-1 in 2001, which replaces ARI 320, 325 and 330. The new ISO standard produces slightly higher ratings because it no longer budgets any electricity for water pumps.

Efficient compressors, variable speed compressors and larger heat exchangers all contribute to heat pump efficiency. Residential ground source heat pumps on the market today have standard COPs ranging from 2.4 to 5.0 and EERs ranging from 10.6 to 30. To qualify for an Energy Star label, heat pumps must meet certain minimum COP and EER ratings which depend on the ground heat exchanger type. For closed loop systems, the ISO 13256-1 heating COP must be 3.3 or greater and the cooling EER must be 14.1 or greater.

Actual installation conditions may produce better or worse efficiency than the standard test conditions. COP improves with a lower temperature difference between the input and output of the heat pump, so the stability of ground temperatures is important. If the loop field or water pump is undersized, the addition or removal of heat may push the ground temperature beyond standard test conditions, and performance will be degraded. Similarly, an undersized blower may allow the plenum coil to overheat and degrade performance.

Soil without artificial heat addition or subtraction and at depths of several metres or more remains at a relatively constant temperature year round. This temperature equates roughly to the average annual air-temperature of the chosen location, usually 7–12 °C (45–54 °F) at a depth of 6 metres (20 ft) in the northern US. Because this temperature remains more constant than the air temperature throughout the seasons, geothermal heat pumps perform with far greater efficiency during extreme air temperatures than air conditioners and air-source heat pumps.

Standards ARI 210 and 240 define Seasonal Energy Efficiency Ratio (SEER) and Heating Seasonal Performance Factors (HSPF) to account for the impact of seasonal variations on air source heat pumps. These numbers are normally not applicable and should not be compared to ground source heat pump ratings. However, Natural Resources Canada has adapted this approach to calculate typical seasonally adjusted HSPFs for ground-source heat pumps in Canada. The NRC HSPFs ranged from 8.7 to 12.8 BTU/hr/watt (2.6 to 3.8 in nondimensional factors, or 255% to 375% seasonal average electricity utilization efficiency) for the most populated regions of Canada. When combined with the thermal efficiency of electricity, this corresponds to net average thermal efficiencies of 100% to 150%.

Environmental Impact

The US Environmental Protection Agency (EPA) has called ground source heat pumps the most energy-efficient, environmentally clean, and cost-effective space conditioning systems available. Heat pumps offer significant emission reductions potential, particularly where they are used for both heating and cooling and where the electricity is produced from renewable resources.

Ground-source heat pumps have unsurpassed thermal efficiencies and produce zero emissions locally, but their electricity supply includes components with high greenhouse gas emissions, unless the owner has opted for a 100% renewable energy supply. Their environmental impact therefore depends on the characteristics of the electricity supply and the available alternatives.

Annual greenhouse gas (GHG) savings from using a ground source heat pump instead of a high-efficiency furnace in a detached residence (assuming no specific supply of renewable energy)				
Country	Electricity CO_2 Emissions Intensity	GHG savings relative to		
		natural gas	heating oil	electric heating
Canada	223 ton/GWh	2.7 ton/yr	5.3 ton/yr	3.4 ton/yr
Russia	351 ton/GWh	1.8 ton/yr	4.4 ton/yr	5.4 ton/yr
US	676 ton/GWh	-0.5 ton/yr	2.2 ton/yr	10.3 ton/yr
China	839 ton/GWh	-1.6 ton/yr	1.0 ton/yr	12.8 ton/yr

The GHG emissions savings from a heat pump over a conventional furnace can be calculated based on the following formula:

- HL = seasonal heat load ≈ 80 GJ/yr for a modern detached house in the northern US

- FI = emissions intensity of fuel = 50 kg(CO_2)/GJ for natural gas, 73 for heating oil, 0 for 100% renewable energy such as wind, hydro, photovoltaic or solar thermal

- AFUE = furnace efficiency ≈ 95% for a modern condensing furnace

- COP = heat pump coefficient of performance ≈ 3.2 seasonally adjusted for northern US heat pump

- EI = emissions intensity of electricity ≈ 200-800 ton(CO_2)/GWh, depending on region

Ground-source heat pumps always produce fewer greenhouse gases than air conditioners, oil furnaces, and electric heating, but natural gas furnaces may be competitive depending on the greenhouse gas intensity of the local electricity supply. In countries like

Canada and Russia with low emitting electricity infrastructure, a residential heat pump may save 5 tons of carbon dioxide per year relative to an oil furnace, or about as much as taking an average passenger car off the road. But in cities like Beijing or Pittsburgh that are highly reliant on coal for electricity production, a heat pump may result in 1 or 2 tons more carbon dioxide emissions than a natural gas furnace. For areas not served by utility natural gas infrastructure, however, no better alternative exists.

The fluids used in closed loops may be designed to be biodegradable and non-toxic, but the refrigerant used in the heat pump cabinet and in direct exchange loops was, until recently, chlorodifluoromethane, which is an ozone depleting substance. Although harmless while contained, leaks and improper end-of-life disposal contribute to enlarging the ozone hole. For new construction, this refrigerant is being phased out in favor of the ozone-friendly but potent greenhouse gas R410A. The EcoCute water heater is an air-source heat pump that uses carbon dioxide as its working fluid instead of chlorofluorocarbons. Open loop systems (i.e. those that draw ground water as opposed to closed loop systems using a borehole heat exchanger) need to be balanced by reinjecting the spent water. This prevents aquifer depletion and the contamination of soil or surface water with brine or other compounds from underground.

Before drilling, the underground geology needs to be understood, and drillers need to be prepared to seal the borehole, including preventing penetration of water between strata. The unfortunate example is a geothermal heating project in Staufen im Breisgau, Germany which seems the cause of considerable damage to historical buildings there. In 2008, the city centre was reported to have risen 12 cm, after initially sinking a few millimeters. The boring tapped a naturally pressurized aquifer, and via the borehole this water entered a layer of anhydrite, which expands when wet as it forms gypsum. The swelling will stop when the anhydrite is fully reacted, and reconstruction of the city center "is not expedient until the uplift ceases." By 2010 sealing of the borehole had not been accomplished. By 2010, some sections of town had risen by 30 cm.

Ground-source heat pump technology, like building orientation, is a natural building technique (bioclimatic building).

Economics

Ground source heat pumps are characterized by high capital costs and low operational costs compared to other HVAC systems. Their overall economic benefit depends primarily on the relative costs of electricity and fuels, which are highly variable over time and across the world. Based on recent prices, ground-source heat pumps currently have lower operational costs than any other conventional heating source almost everywhere in the world. Natural gas is the only fuel with competitive operational costs, and only in a handful of countries where it is exceptionally cheap, or where electricity is exceptionally expensive. In general, a homeowner may save anywhere from 20% to 60% annually on utilities by switching from an ordinary system to a ground-source system.

Capital costs and system lifespan have received much less study until recently, and the return on investment is highly variable. The most recent data from an analysis of 2011-2012 incentive payments in the state of Maryland showed an average cost of residential systems of $1.90 per watt, or about $26,700 for a typical (4 ton) home system. An older study found the total installed cost for a system with 10 kW (3 ton) thermal capacity for a detached rural residence in the US averaged $8000–$9000 in 1995 US dollars. More recent studies found an average cost of $14,000 in 2008 US dollars for the same size system. The US Department of Energy estimates a price of $7500 on its website, last updated in 2008. One source in Canada placed prices in the range of $30,000-$34,000 Canadian dollars. The rapid escalation in system price has been accompanied by rapid improvements in efficiency and reliability. Capital costs are known to benefit from economies of scale, particularly for open loop systems, so they are more cost-effective for larger commercial buildings and harsher climates. The initial cost can be two to five times that of a conventional heating system in most residential applications, new construction or existing. In retrofits, the cost of installation is affected by the size of living area, the home's age, insulation characteristics, the geology of the area, and location of the property. Proper duct system design and mechanical air exchange should be considered in the initial system cost.

Payback period for installing a ground source heat pump in a detached residence			
Country	**Payback period for replacing**		
	natural gas	**heating oil**	**electric heating**
Canada	13 years	3 years	6 years
US	12 years	5 years	4 years
Germany	net loss	8 years	2 years
Notes:			

- Highly variable with energy prices.
- Government subsidies not included.
- Climate differences not evaluated.

Capital costs may be offset by government subsidies; for example, Ontario offered $7000 for residential systems installed in the 2009 fiscal year. Some electric companies offer special rates to customers who install a ground-source heat pump for heating or cooling their building. Where electrical plants have larger loads during summer months and idle capacity in the winter, this increases electrical sales during the winter months. Heat pumps also lower the load peak during the summer due to the increased efficiency of heat pumps, thereby avoiding costly construction of new power plants. For the same reasons, other utility companies have started to pay for the installation of ground-source heat pumps at customer residences. They lease the systems to their customers for a monthly fee, at a net overall saving to the customer.

The lifespan of the system is longer than conventional heating and cooling systems. Good data on system lifespan is not yet available because the technology is too recent, but many early systems are still operational today after 25–30 years with routine maintenance. Most loop fields have warranties for 25 to 50 years and are expected to last at least 50 to 200 years. Ground-source heat pumps use electricity for heating the house. The higher investment above conventional oil, propane or electric systems may be returned in energy savings in 2–10 years for residential systems in the US. If compared to natural gas systems, the payback period can be much longer or non-existent. The payback period for larger commercial systems in the US is 1–5 years, even when compared to natural gas. Additionally, because geothermal heat pumps usually have no outdoor compressors or cooling towers, the risk of vandalism is reduced or eliminated, potentially extending a system's lifespan.

Ground source heat pumps are recognized as one of the most efficient heating and cooling systems on the market. They are often the second-most cost effective solution in extreme climates (after co-generation), despite reductions in thermal efficiency due to ground temperature. (The ground source is warmer in climates that need strong air conditioning, and cooler in climates that need strong heating.)

Commercial systems maintenance costs in the US have historically been between $0.11 to $0.22 per m² per year in 1996 dollars, much less than the average $0.54 per m² per year for conventional HVAC systems.

Governments that promote renewable energy will likely offer incentives for the consumer (residential), or industrial markets. For example, in the United States, incentives are offered both on the state and federal levels of government. In the United Kingdom the Renewable Heat Incentive provides a financial incentive for generation of renewable heat based on metered readings on an annual basis for 20 years for commercial buildings. The domestic Renewable Heat Incentive is due to be introduced in Spring 2014 for seven years and be based on deemed heat.

Installation

Because of the technical knowledge and equipment needed to design and size the system properly (and install the piping if heat fusion is required), a GSHP system installation requires a professional's services. Several installers have published real-time views of system performance in an online community of recent residential installations. The International Ground Source Heat Pump Association (IGSHPA), Geothermal Exchange Organization (GEO), the Canadian GeoExchange Coalition and the Ground Source Heat Pump Association maintain listings of qualified installers in the US, Canada and the UK. Furthermore, detailed analysis of Soil thermal conductivity for horizontal systems and formation thermal conductivity for vertical systems will generally result in more accurately design systems with a higher efficiency.

Seasonal Thermal Energy Storage

Seasonal thermal energy storage (or STES) is the storage of heat or cold for periods of up to several months. The thermal energy can be collected whenever it is available and be used whenever needed, such as in the opposing season. For example, heat from solar collectors or waste heat from air conditioning equipment can be gathered in hot months for space heating use when needed, including during winter months. Waste heat from industrial process can similarly be stored and be used much later. Or the natural cold of winter air can be stored for summertime air conditioning. STES stores can serve district heating systems, as well as single buildings or complexes. Among seasonal storages used for heating, the design peak annual temperatures generally are in the range of 27 to 80 °C (80.6 to 176.0 °F), and the temperature difference occurring in the storage over the course of a year can be several tens of degrees. Some systems use a heat pump to help charge and discharge the storage during part or all of the cycle. For cooling applications, often only circulation pumps are used. A less common term for STES technologies is interseasonal thermal energy storage

Examples for district heating include Drake Landing Solar Community where ground storage provides 97% of yearly consumption without heat pumps, and Danish pond storage with boosting.

STES Technologies

There are several types of STES technology, covering a range of applications from single small buildings to community district heating networks. Generally, efficiency increases and the specific construction cost decreases with size.

Underground Thermal Energy Storage

- UTES (underground thermal energy storage), in which the storage medium may be geological strata ranging from earth or sand to solid bedrock, or aquifers. UTES technologies include:

    o ATES (aquifer thermal energy storage). An ATES store is composed of a doublet, totaling two or more wells into a deep aquifer that is contained between impermeable geological layers above and below. One half of the doublet is for water extraction and the other half for reinjection, so the aquifer is kept in hydrological balance, with no net extraction. The heat (or cold) storage medium is the water and the substrate it occupies. Germany's Reichstag building has been both heated and cooled since 1999 with ATES stores, in two aquifers at different depths. In the Netherlands there are now well over 1,000 ATES systems, which are now a standard construction option. A significant system has been operating at Richard Stockton College (New Jersey) for several years. ATES has a lower installation cost than BTES because usually fewer holes are drilled, but ATES has a higher operating cost. Also, ATES requires particular underground conditions to be feasible, including the presence of an aquifer.

o   BTES (borehole thermal energy storage). BTES stores can be constructed wherever boreholes can be drilled, and are composed of one to hundreds of vertical boreholes, typically 155 mm (6.102 in) in diameter. Systems of all sizes have been built, including many quite large. The strata can be anything from sand to crystalline hardrock, and depending on engineering factors the depth can be from 50 to 300 metres (164 to 984 ft). Spacings have ranged from 3 to 8 metres (9.8 to 26.2 ft). Thermal models can be used to predict seasonal temperature variation in the ground, including the establishment of a stable temperature regime which is achieved by matching the inputs and outputs of heat over one or more annual cycles. Warm-temperature seasonal heat stores can be created using borehole fields to store surplus heat captured in summer to actively raise the temperature of large thermal banks of soil so that heat can be extracted more easily (and more cheaply) in winter. Interseasonal Heat Transfer uses water circulating in pipes embedded in asphalt solar collectors to transfer heat to Thermal Banks created in borehole fields. A ground source heat pump is used in winter to extract the warmth from the Thermal Bank to provide space heating via underfloor heating. A high Coefficient of Performance is obtained because the heat pump starts with a warm temperature of 25 °C (77 °F) from the thermal store, instead of a cold temperature of 10 °C (50 °F) from the ground. A BTES operating at Richard Stockton College since 1995 at a peak of about 29 °C (84.2 °F) consists of 400 boreholes 130 metres (427 ft) deep under a 3.5-acre (1.4 ha) parking lot. It has a heat loss of 2% over six months. The upper temperature limit for a BTES store is 85 °C (185 °F) due to characteristics of the PEX pipe used for BHEs, but most do not approach that limit. Boreholes can be either grout- or water-filled depending on geological conditions, and usually have a life expectancy in excess of 100 years. Both a BTES and its associated district heating system can be expanded incrementally after operation begins, as at Neckarsulm, Germany. BTES stores generally do not impair use of the land, and can exist under buildings, agricultural fields and parking lots. An example of one of the several kinds of STES illustrates well the capability of interseasonal heat storage. In Alberta, Canada, the homes of the Drake Landing Solar Community (in operation since 2007), get 97% of their year-round heat from a district heat system that is supplied by solar heat from solar-thermal panels on garage roofs. This feat – a world record – is enabled by interseasonal heat storage in a large mass of native rock that is under a central park. The thermal exchange occurs via a cluster of 144 boreholes, drilled 37 metres (121 ft) into the earth. Each borehole is 155 mm (6.1 in) in diameter and contains a simple heat exchanger made of small diameter plastic pipe, through which water is circulated. No heat pumps are involved.

o   Exchange (cavern or mine thermal energy storage). STES stores are possible in flooded mines, purpose-built chambers, or abandoned underground oil stores (e.g. those mined into crystalline hardrock in Norway), if they are close enough to a heat (or cold) source and market.

o   Exchange . During construction of large buildings, BHE heat exchangers much like those used for BTES stores have been spiraled inside the cages of reinforcement bars for pilings, with concrete then poured in place. The pilings and surrounding strata then become the storage medium.

o   Exchange (geo interseasonal insulated thermal storage). During construction of any building with a primary slab floor, an area approximately the footprint of the building to be heated, and >1M in depth, is insulated on all 6 sides typically with HDPE closed cell insulation. Pipes are used to transfer solar energy into the insulated area, as well as extracting heat as required on demand. If there is significant internal ground water flow, remedial actions are needed to prevent it.

Surface and Above Ground Technologies

*   Exchange . Lined, shallow dug pits that are filled with gravel and water as the storage medium are used for STES in many Danish district heating systems. Storages pits are covered with a layer of insulation and then soil, and are used for agriculture or other purposes. Marstal, Denmark's system is a case study, initially providing 20% of the village's year-round heat but now being expanded to provide twice that. The worlds largest pit store (200,000 m3) was commissioned in Denmark in 2015, and allows solar heat to provide 50% of the annual energy for the world's largest solar-enabled district heating system.

*   Exchange . Large scale STES water storage tanks can be built above ground, insulated, and then covered with soil.

*   Exchange . For small installations, a heat exchanger of corrugated plastic pipe can be shallow-buried in a trench to create a STES.

*   Exchange , with passive heat storage in surrounding soil.

Conferences and Organizations

The International Energy Agency's *Energy Conservation through Energy Storage (ECES) Programme* has held triennial global energy conferences since 1981. The conferences originally focused exclusively on STES, but now that those technologies are mature other topics such as phase change materials (PCM) and electrical energy storage are also being covered. Since 1985 each conference has had "stock" (for storage) at the end of its name; e.g. Ecostock, Thermastock. They are held at various locations around the world. Most recent was Innostock 2012 (the 12th International Conference on Thermal Energy Storage) in Lleida, Spain. Greenstock 2015 will be held in Beijing.

The IEA-ECES programme continues the work of the earlier *International Council for Thermal Energy Storage* which from 1978 to 1990 had a quarterly newsletter and was initially sponsored by the U.S. Department of Energy. The newsletter was initially called *ATES Newsletter,* and after BTES became a feasible technology it was changed to *STES Newsletter.*

Use of STES for Small, Passively Heated Buildings

Small passively heated buildings typically use the soil adjoining the building as a low-temperature seasonal heat store that in the annual cycle reaches a maximum temperature similar to average annual air temperature, with the temperature drawn down for heating in colder months. Such systems are a feature of building design, as some simple but significant differences from 'traditional' buildings are necessary. At a depth of about 20 feet (6.1 m) in the soil, the temperature is naturally stable within a year-round range, if the draw down does not exceed the natural capacity for solar restoration of heat. Such storage systems operate within a narrow range of storage temperatures over the course of a year, as opposed to the other STES systems described above for which large annual temperature differences are intended.

Two basic passive solar building technologies were developed in the US during the 1970s and 1980s. They utilize direct heat conduction to and from thermally isolated, moisture-protected soil as a seasonal storage medium for space heating, with direct conduction as the heat return method. In one method, "passive annual heat storage" (PAHS), the building's windows and other exterior surfaces capture solar heat which is transferred by conduction through the floors, walls, and sometimes the roof, into adjoining thermally buffered soil.

When the interior spaces are cooler than the storage medium, heat is conducted back to the living space. The other method, "annualized geothermal solar" (AGS) uses a separate solar collector to capture heat. The collected heat is delivered to a storage device (soil, gravel bed or water tank) either passively by the convection of the heat transfer medium (e.g. air or water) or actively by pumping it. This method is usually implemented with a capacity designed for six months of heating.

A number of examples of the use of solar thermal storage from across the world include: Suffolk One a college in East Anglia, England that uses a thermal collector of pipe buried in the bus turning area to collect solar energy that is then stored in 18 100 metres (330 ft) boreholes for use in winter heating. Drake Landing Solar Community in Canada uses solar thermal collectors on the garage roofs of 52 homes, which is then stored in an array of 35 metres (115 ft) deep boreholes. The ground can reach temperatures in excess of 70 °C which is then used to heat the houses passively. The scheme has been running successfully since 2007. In Brædstrup, Denmark some 8,000 square metres (86,000 sq ft) of solar thermal collectors are used to collect some 4,000,000 kWh/a similarly stored in an array of 50 metres (160 ft) deep boreholes.

Liquid Engineering

Architect Matyas Gutai obtained an EU grant to construct a house in Hungary which uses extensive water filled wall panels as heat collectors and reservoirs with underground heat storage water tanks. The design uses microprocessor control.

Small Buildings with Internal STES Water Tanks

A number of homes and small apartment buildings have demonstrated combining a large internal water tank for heat storage with roof-mounted solar-thermal collectors. Storage temperatures of 90 °C (194 °F) are sufficient to supply both domestic hot water and space heating. The first such house was MIT Solar House #1, in 1939. An eight-unit apartment building in Oberburg, Switzerland was built in 1989, with three tanks storing a total of 118 m³ (4,167 cubic feet) that store more heat than the building requires. Since 2011, that design is now being replicated in new buildings.

In Berlin, the "Zero Heating Energy House", was built in 1997 in as part of the IEA Task 13 low energy housing demonstration project. It stores water at temperatures up to 90 °C (194 °F) inside a 20 m³ (706 cubic feet) tank in the basement,.

A similar example was built in Ireland in 2009, as a prototype. The *solar seasonal store* consists of a 23 m³ (812 cu ft) tank, filled with water, which was installed in the ground, heavily insulated all around, to store heat from evacuated solar tubes during the year. The system was installed as an experiment to heat the *world's first standardized pre-fabricated passive house* in Galway, Ireland. The aim was to find out if this heat would be sufficient to eliminate the need for any electricity in the already highly efficient home during the winter months.

Use of STES in Greenhouses

STES is also used extensively for the heating of greenhouses. ATES is the kind of storage commonly in use for this application. In summer, the greenhouse is cooled with ground water, pumped from the "cold well" in the aquifer. The water is heated in the process, and is returned to the "warm well" in the aquifer. When the greenhouse needs heat, such as to extend the growing season, water is withdrawn from the warm well, becomes chilled while serving its heating function, and is returned to the cold well. This is a very efficient system of free cooling, which uses only circulation pumps and no heat pumps.

References

- "annex 9". National Inventory Report 1990–2006:Greenhouse Gas Sources and Sinks in Canada. Government of Canada. May 2008. ISBN 978-1-100-11176-6. ISSN 1706-3353.

- Hestnes, A.; Hastings, R. (eds) (2003). Solar Energy Houses: Strategies, Technologies, Examples. pp.109-114. ISBN 1-902916-43-3.

- "OpenThermal.org analysis of geothermal incentive payments in the state of Maryland". OpenThermal.org. Retrieved 17 May 2015.

- "White House Executive Order on Sustainability Includes Geothermal Heat Pumps". www. geoexchange.org. Retrieved 17 May 2015.

- SDH (Solar District Heating) Newsletter (2014). The world's largest solar heating plant to be established in Vojens, Denmark. 7 June 2014.

- "Environmental Technology Verification Report" (PDF). U.S. Environmental Protection Agency. Archived from the original (PDF) on 2014-04-19. Retrieved December 3, 2015.

- Wong, Bill (June 28, 2011), "Drake Landing Solar Community" (PDF), Drake Landing Solar Community, IDEA/CDEA District Energy/CHP 2011 Conference, Toronto, pp. 1–30, retrieved 21 April 2013

- "Geothermal Technologies Program: Tennessee Energy Efficient Schools Initiative Ground Source Heat Pumps". Apps1.eere.energy.gov. 2010-03-29. Retrieved 2011-03-30.

Air Quality Laws

Air quality law is the law that governs the release of air pollutants into the air. It helps in the regulation of the quality of air inside and around building. The aspects elucidates in this chapter are of vital importance and provides a better understanding of air quality law.

Air Quality Law

Exchange govern the emission of air pollutants into the atmosphere. A specialized subset of air quality laws regulate the quality of air inside buildings. Air quality laws are often designed specifically to protect human health by limiting or eliminating airborne pollutant concentrations. Other initiatives are designed to address broader ecological problems, such as limitations on chemicals that affect the ozone layer, and emissions trading programs to address acid rain or climate change. Regulatory efforts include identifying and categorizing air pollutants, setting limits on acceptable emissions levels, and dictating necessary or appropriate mitigation technologies.

Air Pollutant Classification

Air quality regulation must identify the substances and energies which qualify as "pollution" for purposes of further control. While specific labels vary from jurisdiction to jurisdiction, there is broad consensus among many governments regarding what constitutes air pollution. For example, the United States Clean Air Act identifies ozone, particulate matter, carbon monoxide, nitrogen oxides (NO_x), sulfur dioxide (SO_2), and lead (Pb) as "criteria" pollutants requiring nationwide regulation. EPA has also identified over 180 compounds it has classified as "hazardous" pollutants requiring strict control. Other compounds have been identified as air pollutants due to their adverse impact on the environment (e.g., CFCs as agents of ozone depletion), and on human health (e.g., asbestos in indoor air). A broader conception of air pollution may also incorporate noise, light, and radiation. The United States has recently seen controversy over whether carbon dioxide (CO_2) and other greenhouse gases should be classified as air pollutants.

Air Quality Standards

Air quality standards are legal standards or requirements governing concentrations of air pollutants in breathed air, both outdoors and indoors. Such standards generally are expressed as levels of specific air pollutants that are deemed acceptable in ambient air, and are most often designed to reduce or eliminate the human health effects of air pollution, although secondary effects such as crop and building damage may also be considered. Determining appropriate air quality standards generally requires up-to-date scientific data on the health effects of the pollutant under review, with specific information on exposure times and sensitive populations. It also generally requires periodic or continuous monitoring of air quality.

As an example, the United States Environmental Protection Agency has developed the National Ambient Air Quality Standards (NAAQS) NAAQS set attainment thresholds for sulfur dioxide, particulate matter (PM_{10} and $PM_{2.5}$), carbon monoxide, ozone, nitrogen oxides NO_x, and lead (Pb) in outdoor air throughout the United States. Another set of standards, for indoor air in employment settings, is administered by the U.S. Occupational Safety and Health Administration.

A distinction may be made between mandatory and aspirational air quality standards. For example, U.S. state governments must work toward achieving NAAQS, but are not forced to meet them. On the other hand, employers may be required immediately to rectify any violation of OSHA workplace air quality standards.

Emission Standards

Emission standards are the legal requirements governing air pollutants released into the atmosphere. Emission standards set quantitative limits on the permissible amount of specific air pollutants that may be released from specific sources over specific timeframes. They are generally designed to achieve air quality standards and to protect human health.

Numerous methods exist for determining appropriate emissions standards, and different regulatory approaches may be taken depending on the source, industry, and air pollutant under review. Specific limits may be set by reference to and within the confines of more general air quality standards. Specific sources may be regulated by means of performance standards, meaning numerical limits on the emission of a specific pollutant from that source category. Regulators may also mandate the adoption and use of specific control technologies, often with reference to feasibility, availability, and cost. Still other standards may be set using performance as a benchmark - for example, requiring all of a specific type of facility to meet the emissions limits achieved by the best performing facility of the group. All of these methods may be modified by incorporating emissions averaging, market mechanisms such as emissions trading, and other alternatives.

For example, all of these approaches are used in the United States. The United States Environmental Protection Agency (responsible for air quality regulation at a national level under the U.S. Clean Air Act, utilizes performance standards under the New Source Performance Standard (NSPS) program. Technology requirements are set under RACT (Reasonably Available Control Technology), BACT (Best Available Control Technology), and LAER (Lowest Achievable Emission Rate) standards. Flexibility alternatives are implemented in U.S. programs to eliminate acid rain, protect the ozone layer, achieve permitting standards, and reduce greenhouse gas emissions.

Control Technology Requirements

In place of or in combination with air quality standards and emission control standards, governments may choose to reduce air pollution by requiring regulated parties to adopt emissions control technologies (i.e., technology that reduces or eliminates emissions). Such devices include but are not limited to flare stacks, incinerators, catalytic combustion reactors, selective catalytic reduction reactors, electrostatic precipitators, baghouses, wet scrubbers, cyclones, thermal oxidizers, Venturi scrubbers, carbon adsorbers, and biofilters.

The selection of emissions control technology may be the subject of complex regulation that may balance multiple conflicting considerations and interests, including economic cost, availability, feasibility, and effectiveness. The various weight given to each factor may ultimately determine the technology selected. The outcome of an analysis seeking a technology that all players in an industry can afford could be different from an analysis seeking to require all players to adopt the most effective technology yet developed, regardless of cost. For example, the United States Clean Air Act contains several control technology requirements, including Best Available Control Technology (BACT) (used in New Source Review), Reasonably Available Control Technology (RACT) (existing sources), Lowest Achievable Emissions Rate (LAER) (used for major new sources in non-attainment areas), and Maximum Achievable Control Technology (MACT) standards.

Bans

Air quality laws may take the form of bans. While arguably a class of emissions control law (where the emission limit is set to zero), bans differ in that they may regulate activity other than the emission of a pollutant itself, even though the ultimate goal is to eliminate the emission of the pollutant.

A common example is a burn ban. Residential and commercial burning of wood materials may be restricted during times of poor air quality, eliminating the immediate emission of particulate matter and requiring use of non-polluting heating methods. A more significant example is the widespread ban on the manufacture of dichlorodifluoromethane (Freon)), formerly the standard refrigerant in automobile air conditioning systems. This substance, often released into the atmosphere unintentionally as a result of refrigerant system leaks, was determined to have a significant ozone depletion potential, and its widespread use to pose a significant threat to the Earth's ozone layer. Its manufacture was prohibited as part of a suite of restrictions adopted internationally in the Montreal Protocol to the Vienna Convention for the Protection of the Ozone Layer. Still another example is the ban on use of asbestos in building construction materials, to eliminate future exposure to carcinogenic asbestos fibers when the building materials are disturbed.

Data Collection and Access

Air quality laws may impose substantial requirements for collecting, storing, submitting, and providing access to technical data for various purposes, including regulatory enforcement, public health programs, and policy development.

Data collection processes may include monitoring ambient air for the presence of pollutants, directly monitoring emissions sources, or collecting other quantitative information from which air quality information may be deduced. For example, local agencies may employ a particulate matter sampler to determine ambient air quality in a locality over time. Fossil power plants may required to monitor emissions at a flue-gas stack to determine quantities of relevant pollutants emitted. Automobile manufacturers may be required to collect data regarding car sales, which, when combined with technical specifications regarding fuel consumption and efficiency, may be used to estimate total vehicle emissions. In each case, data collection may be short- or long-term, and at varying frequency (e.g., hourly, daily).

Air quality laws may include detailed requirements for recording, storing, and submitting relevant information, generally with the ultimate goal of standardizing data practices in order to facilitate data access and manipulation at a later time. Precise requirements may be very difficult to determine without technical training and may change over time in response to, for example, changes in law, changes in policy, changes in available technology, and changes in industry practice. Such requirements may be developed at a national level and reflect consensus or compromise between government agencies, regulated industry, and public interest groups.

Once air quality data are collected and submitted, some air quality laws may require government agencies or private parties to provide the public with access to the information - whether the raw data alone, or via tools to make the data more useful, accessible, and understandable. Where public access mandates are general, it may be left to the collecting agency to decide whether and to what extent the data is to be centralized and organized. For example, the United States Environmental Protection Agency, National Oceanic and Atmospheric Administration, National Park Service, and tribal, state, and local agencies coordinate to produce an online mapping and data access tool called AirNow, which provides real-time public access to U.S. air quality index information, searchable by location.

Once data are collected and published, they may be used as inputs in mathematical models and forecasts. For example, atmospheric dispersion modeling may be employed to examine the potential impact of new regulatory requirements on existing populations or geographic areas. Such models in turn could drive changes in data collection and reporting requirements.

Controversy

Proponents of air quality law argue that they have caused or contributed to major reductions in air pollution, with concomitant human health and environmental benefits, even in the face of large-scale economic growth and increases in motor vehicle use. On the other hand, controversy may arise over the estimated cost of additional regulatory standards.

Arguments over cost, however, cut both ways. For example, the "estimates that the benefits of reducing fine particle and ground level ozone pollution under the 1990 Clean Air Act amendments will reach approximately $2 trillion in 2020 while saving 230,000 people from early death in that year alone." According to the same report, 2010 alone the reduction of ozone and particulate matter in the atmosphere prevented more than 160,000 cases of premature mortality, 130,000 heart attacks, 13 million lost work days and 1.7 million asthma attacks. Criticisms of EPA's methodologies in reaching these and similar numbers are publicly available.

Around the World

International Law

International law includes agreements related to trans-national air quality, including greenhouse gas emissions:

- Convention on Long-Range Transboundary Air Pollution (LRTAP), Geneva, 1979

- Environmental Protection: Aircraft Engine Emissions, Annex 16, vol. 2 to the Chicago Convention on International Civil Aviation, Montreal, 1981

- Framework Convention on Climate Change (UNFCCC), New York, 1992, including the Kyoto Protocol, 1997, and the Paris Agreement, 2015

- Georgia Basin-Puget Sound International Airshed Strategy, Vancouver, Statement of Intent, 2002

- U.S.-Canada Air Quality Agreement (bilateral U.S.-Canadian agreement on acid rain), 1986

- Vienna Convention for the Protection of the Ozone Layer, Vienna, 1985, including the Montreal Protocol on Substances that Deplete the Ozone Layer, Montreal 1987

Canada

With some industry-specific exceptions, Canadian air pollution regulation was traditionally handled at the provincial level. However, under the authority of the Canadian Environmental Protection Act, 1999, the country has recently enacted a national program called the Canadian Air Quality Management System (AQMS). The program includes five main regulatory mechanisms: the Canadian Ambient Air Quality Standards (CAAQS); Base Level Industrial Emission Requirements (BLIERs) (emissions controls and technology); management of local air quality through the management of Local Air Zones; management of regional air quality through the management of Regional Airsheds; and collaboration to reduce mobile source emissions.

The Canadian government has also made efforts to pass legislation related to the country's greenhouse gas emissions. It has passed laws related to fuel economy in passenger vehicles and light trucks, heavy-duty vehicles, renewable fuels, and the energy and transportation sectors.

China

China, with severe air pollution in mega-cities and industrial centers, particularly in the north, has adapted the Airborne Pollution Prevention and Control Action Plan which aims for a 25% reduction in air pollution by 2017 from 2012 levels. Funded by $277 billion from the central government, the action plan targets PM 2.5 particulates which affect human health.

New Zealand

New Zealand passed its Clean Air Act 1972 in response to growing concerns over industrial and urban air pollution. That Act classified sources, imposed permitting requirements, and created a process for determining requisite control technology. Local authorities were authorized to regulate smaller polluters. Within the Christchurch Clean Air Zone, burn bans and other measures were effected to control smog.

The Clean Air Act 1972 was replaced by the Resource Management Act 1991. The act did not set air quality standards, but did provide for national guidance to be developed. This resulted in the promulgation of New Zealand's National Environmental Standards for Air Quality in 2004 with subsequent amendments.

United Kingdom

In response to the Great Smog of 1952, the British Parliament introduced the Clean Air Act 1956. This act legislated for zones where smokeless fuels had to be burnt and relocated power stations to rural areas. The Clean Air Act 1968 introduced the use of tall chimneys to disperse air pollution for industries burning coal, liquid or gaseous fuels. The Clean Air Act was updated in 1993 and can be reviewed online legislation

Clean Air Act 1993. The biggest domestic impact comes from Part III, Smoke Control Areas, which are designated by local authorities and can vary by street in large towns.

United States

The primary law regulating air quality in the United States is the U.S. Clean Air Act. The law was initially enacted as the Air Pollution Control Act of 1955. Amendments in 1967 and 1970 (the framework for today's U.S. Clean Air Act) imposed national air quality requirements, and placed administrative responsibility with the newly created Environmental Protection Agency. Major amendments followed in 1977 and 1990. State and local governments have enacted similar legislation, either implementing federal programs or filling in locally important gaps in federal programs.

Clean Air Act (United States)

The Exchange is a United States federal law designed to control air pollution on a national level. It is one of the United States' first and most influential modern environmental laws, and one of the most comprehensive air quality laws in the world. As with many other major U.S. federal environmental statutes, it is administered by the U.S. Environmental Protection Agency (EPA), in coordination with state, local, and tribal governments. Its implementing regulations are codified at 40 C.F.R. Subchapter C, Parts 50-97.

The 1955 Air Pollution Control Act was the first U.S federal legislation that pertained to air pollution; it also provided funds for federal government research of air pollution. The first federal legislation to actually pertain to "*controlling*" air pollution was the Clean Air Act of 1963. The 1963 act accomplished this by establishing a federal program within the U.S. Public Health Service and authorizing research into techniques for monitoring and controlling air pollution. In 1967, the Air Quality Act enabled the federal government to increase its activities to investigate enforcing interstate air pollution transport, and, for the first time, to perform far-reaching ambient monitoring studies and stationary source inspections. The 1967 act also authorized expanded studies of air pollutant emission inventories, ambient monitoring techniques, and control techniques.

Major amendments to the law, requiring regulatory controls for air pollution, passed in 1970, 1977 and 1990.

The 1970 amendments greatly expanded the federal mandate, requiring comprehensive federal and state regulations for both stationary (industrial) pollution sources and mobile sources. It also significantly expanded federal enforcement. Also, the Environmental Protection Agency was established on December 2, 1970 for the purpose of consolidating pertinent federal research, monitoring, standard-setting and enforcement activities into one agency that ensures environmental protection.

The 1990 amendments addressed acid rain, ozone depletion, and toxic air pollution, established a national permits program for stationary sources, and increased enforcement authority. The amendments also established new auto gasoline reformulation requirements, set Reid vapor pressure (RVP) standards to control evaporative emissions from gasoline, and mandated new gasoline formulations sold from May to September in many states.

Reviewing his tenure as EPA Administrator under President George H. Bush, William K. Reilly characterized passage of the 1990 Clean Air Act as his most notable accomplishment.

The Clean Air Act was the first major environmental law in the United States to include a provision for citizen suits. Numerous state and local governments have enacted similar legislation, either implementing federal programs or filling in locally important gaps in federal programs.

Components of Air Pollution Prevention and Control

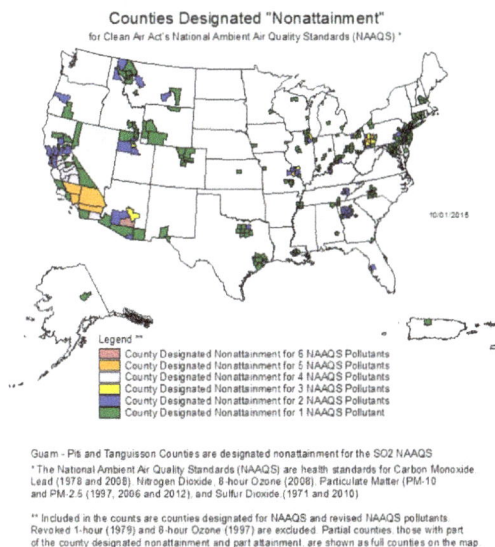

Counties in the United States where one or more Exchange are not met, as of October 2015.

Title I - Programs and Activities

Part A - Air Quality and Emissions Limitations

This section of the act declares that protecting and enhancing the nation's air quality promotes public health. The law encourages prevention of regional air pollution and control programs. It also provides technical and financial assistance for air pollution prevention at both state and local governments. Additional subchapters cover of cooperation, research, investigation, training and other activities. Grants for air pollution planning and control programs, and interstate air quality agencies and program cost limitations are also included in this section of the act.

The act mandates air quality control regions, designated as attainment vs non-attainment. Non-attainment areas do not meet national standards for primary or secondary ambient air quality. Attainment areas meet these standards, while unclassifiable areas cannot be classified on the basis of the information that is available.

Air quality criteria, national primary and secondary ambient air quality standards, state implementation plans and performance standards for new stationary sources are also covered in Part A. The list of hazardous air pollutants established by the act includes acetaldehyde, benzene, chloroform, phenols and selenium compounds. The list also includes mineral fiber emissions from manufacturing or processing glass, rock or slag fibers as well as radioactive atoms. The list periodically can be modified. The act lists unregulated radioactive pollutants such as cadmium, arsenic, and polycyclic organic matter and mandates listing them if they will cause or contribute to air pollution that endangers public health, under section 7408 or 7412.

The remaining subchapters cover smokestack heights, state plan adequacy, and estimating emissions of carbon monoxide, volatile organic compounds, and oxides of nitrogen from area and mobile sources. Measures to prevent unemployment or other economic disruption include using local coal or coal derivatives to comply with implementation requirements. The final subchapter in this act focuses on land use authority.

Part B - Ozone Protection

Because of advances in the atmospheric chemistry, this section was replaced by Title VI when the law was amended in 1990.

This change in the law reflected significant changes in scientific understanding of ozone formation and depletion. Ozone absorbs UVC light and shorter wave UVB, and lets through UVA, which is largely harmless to people. Ozone exists naturally in the stratosphere, not the troposphere. It is laterally distributed because it is destroyed by strong sunlight, so there is more ozone at the poles. Ozone is created when O_2 comes in contact with photons from solar radiation. Therefore, a decrease in the intensity of solar radiation also results in a decrease in the formation of ozone in the stratosphere. This exchange is known as the Chapman mechanism:

O_2 + UV photon → 2 O (note that atmospheric oxygen as O is highly unstable)

$$O + O_2 + M \rightarrow O_3 (O_3 \text{ is Ozone}) + M$$

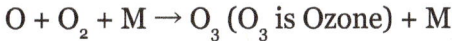

M represents a third molecule, needed to carry off the excess energy of the collision of $O + O_2$.

Atmospheric freon and chlorofluorocarbons (CFCs) contribute to ozone depletion (Chlorine is a catalytic agent in ozone destruction). Following discovery of the ozone hole in 1985, the 1987 Montreal Protocol successfully implemented a plan to replace CFCs and was viewed by some environmentalists as an example of what is possible for the future of environmental issues, if the political will is present.

Part C - Prevention of Significant Deterioration of Air Quality

The Clean Air Act requires permits to build or add to major stationary sources of air pollution. This permitting process, known as New Source Review (NSR), applies to sources in areas that meet air quality standards as well as areas that are unclassifiable. Permits in attainment or unclassifiable areas are referred to as Prevention of Significant Deterioration (PSD) permits, while permits for sources located in nonattainment areas are referred to as nonattainment area (NAA) permits.

The fundamental goals of the PSD program are to:

1. prevent new non-attainment areas by ensuring economic growth in harmony with existing clean air;

2. protect public health and welfare from any adverse effects;

3. preserve and enhance the air quality in national parks and other areas of special natural recreational, scenic, or historic value.

Part D - Plan Requirements for Non-attainment Areas

Under the Clean Air Act states are required to submit a plan for non-attainment areas to reach attainment status as soon as possible but in no more than five years, based on the severity of the air pollution and the difficulty posed by obtaining cleaner air.

The plan must include:

*   an inventory of all pollutants

*   permits

*   control measures, means and techniques to reach standard qualifications

*   contingency measures

The plan must be approved or revised if required for approval, and specify whether local governments or the state will implement and enforce the various changes. Achieving attainment status makes a request for reevaluation possible. It must include a plan for maintenance of air quality.

Title II - Emission Standards for Moving Sources

Part A - Motor Vehicle Emission and Fuel Standards (CAA § 201-219; USC § 7521-7554)

Subchapters of Title II cover state standards and grants, prohibited acts and actions to restrain violations, as well as a study of emissions from nonroad vehicles (other than locomotives) to determine whether they cause or contribute to air pollution. Motorcycles are treated in the same way as automobiles under the emission standards for new motor vehicles or motor vehicle engines. The last few subchapters deal with high altitude performance adjustments, motor vehicle compliance program fees, prohibition on production of engines requiring leaded gasoline and urban bus standards.

This part of the bill was extremely controversial the time it was passed. The automobile industry argued that it could not meet the new standards. Senators expressed concern about impact on the economy. Specific new emissions standards for moving sources passed years later.

Part B - Aircraft Emission Standards

Many volatile organic compounds (VOCs) are emitted over airports and affect the air quality in the region. VOCs include benzene, formaldehyde and butadienes which are known to cause health problems such as birth defects, cancer and skin irritation. Hundreds of tons of emissions from aircraft, ground support equipment, heating systems, and shuttles and passenger vehicles are released into the air, causing smog. Therefore, major cities such as Seattle, Denver, and San Francisco require a Climate Action Plan as well as a greenhouse gas inventory. Additionally, federal programs such as VALE are working to offset costs for programs that reduce emissions.

Title II sets emission standards for airlines and aircraft engines and adopts standards set by the International Civil Aviation Organization (ICAO). However aircraft carbon dioxide emission standards have not been established by either ICAO nor the EPA. It is the responsibility of the Secretary of Transportation, after consultation with the Administrator, to prescribe regulations that comply with section 7571 and ensure the necessary inspections take place.

Part C - Clean Fuel Vehicles

Trucks and automobiles play a large role in deleterious air quality. Harmful chemicals such as nitrogen oxide, hydrocarbons, carbon monoxide and sulfur dioxide are released from motor vehicles. Some of these also react with sunlight to produce photochemicals. These harmful substances change the climate, alter ocean pH and include toxins that may cause cancer, birth defects or respiratory illness. Motor vehicles increased in the 1990s since approximately 58 percent of households owned two or more vehicles. The Clean Fuel Vehicle programs focused on alternative fuel use and petroleum fuels that met low emission vehicle (LEV) levels. Compressed natural gas, ethanol, methanol, liquefied petroleum gas and electricity are examples of cleaner alternative fuel. Programs such as the California Clean Fuels Program and pilot program are increasing demand that for new fuels to be developed to reduce harmful emissions.

The California pilot program incorporated under this section focuses on pollution control in ozone non-attainment areas. The provisions apply to light-duty trucks and light-duty vehicles in California. The state also requires that clean alternative fuels for sale at numerous locations with sufficient geographic distribution for convenience. Production of clean-fuel vehicles isn't mandated except as part of the California pilot program.

Title III - General Provisions

Under the law prior to 1990, EPA was required to construct a list of Hazardous Air Pollutants as well as health-based standards for each one. There were 188 air pollutants listed and the source from which they came. The EPA was given ten years to generate technology-based emission standards. Title III is considered a second phase, allowing the EPA to assess lingering risks after the enactment of the first phase of emission standards. Title III also enacts new standards with regard to the protection of public health.

A citizen may file a lawsuit to obtain compliance with an emission standard issued by the EPA or by a state, unless there is an ongoing enforcement action being pursued by EPA or the appropriate state agency.

Title IV - Noise Pollution

This title pre-dates the *Clean Air Act*. With the passage of the *Clean Air Act*, it became codified as Title IV. However, another Title IV was enacted in the 1970 amendments. The second Title IV was then appended to this Title IV as Title IV-A.

This title established the EPA Office of Noise Abatement and Control to reduce noise pollution in urban areas, to minimize noise-related impacts on psychological and physiological effects on humans, effects on wildlife and property (including values), and other noise-related issues. The agency was also assigned to run experiments to study the effects of noise.

Title IV-A - Acid Deposition Control

This title was added as part of the 1990 amendments. It addresses the issue of acid rain, which is caused by nitrogen oxides (NO_x) and sulfur dioxide (SO_2) emissions from electric power plants powered by fossil fuels, and other industrial sources. The 1990 amendments gave industries more pollution control options including switching to low-sulfur coal and/or adding devices that controlled the harmful emissions. In some cases plants had to be closed down to prevent the dangerous chemicals from entering the atmosphere.

Title IV-A mandated a two-step process to reduce SO_2 emissions. The first stage required more than 100 electric generating facilities larger than 100 megawatts to meet a 3.5 million ton SO_2 emission reduction by January 1995. The second stage gave facilities larger than 75 megawatts a January 2000 deadline.

Title V - Permits

The 1990 amendments authorized a national operating permit program, covering thousands of large industrial and commercial sources. It required large businesses to address pollutants released into the air, measure their quantity, and have a plan to control and minimize them as well as to periodically report. This consolidated requirements for a facility into a single document.

In non-attainment areas, permits were required for sources that emit as little as 50, 25, or 10 tons per year of VOCs depending on the severity of the region's non-attainment status.

Most permits are issued by state and local agencies. If the state does not adequately monitor requirements, the EPA may take control. The public may request to view the permits by contacting the EPA. The permit is limited to no more than five years and requires a renewal.

Title VI - Stratospheric Ozone Protection

Starting in 1990, Title VI mandated regulations regarding the use and production of chemicals that harm the Earth's stratospheric ozone layer. This ozone layer protects against harmful ultraviolet B sunlight linked to several medical conditions including cataracts and skin cancer.

The ozone-destroying chemicals were classified into two groups, Class I and Class II. Class I consists of substances, including chlorofluorocarbons, that have an ozone depletion potential (ODP) (HL) of 0.2 or higher. Class II lists substances, including hydrochlorofluorocarbons, that are known to or may be detrimental to the stratosphere. Both groups have a timeline for phase-out:

- For Class I substances, no more than seven years after being added to the list and

- For Class II substances no more than ten years.

Title VI establishes methods for preventing harmful chemicals from entering the stratosphere in the first place, including recycling or proper disposal of chemicals and finding substitutes that cause less or no damage. The Significant New Alternatives Policy (SNAP) Program is EPA's program to evaluate and regulate substitutes for the ozone-depleting chemicals that are being phased out under the stratospheric ozone protection provisions of the Clean Air Act.

Over 190 countries signed the Montreal Protocol in 1987, agreeing to work to eliminate or limit the use of chemicals with ozone-destroying properties.

History

Legislation

Congress passed the first legislation to address air pollution with the 1955 Air Pollution Control Act that provided funds to the U.S. Public Health service, but did not formulate pollution regulation. However, the Clean Air Act in 1963, created a research and regulatory program in the U.S. Public Health Service. The Act authorized development of emission standards for stationary sources, but not mobile sources of air pollution. The 1967 Exchange mandated enforcement of interstate air pollution standards and authorized ambient monitoring studies and stationary source inspections.

In the Clean Air Act Extension of 1970, Congress greatly expanded the federal mandate by requiring comprehensive federal and state regulations for both industrial and mobile sources. The law established four new regulatory programs:

- National Ambient Air Quality Standards (NAAQS). EPA was required to promulgate national standards for six criteria pollutants: carbon monoxide, nitrogen dioxide, sulfur dioxide, particulate matter, hydrocarbons and photochemical oxidants. (Some of the criteria pollutants were revised in subsequent legislation.)

- State Implementation Plans (SIPs)

- New Source Performance Standards (NSPS); and

- National Emissions Standards for Hazardous Air Pollutants (NESHAPs).

The 1970 law is sometimes called the "Muskie Act" because of the central role Maine Senator Edmund Muskie played in drafting the bill.

To implement the strict new Clean Air Act of 1970, during his first term as EPA Administrator William Ruckelshaus spent 60% of his time on the automobile industry, whose emissions were to be reduced 90% under the new law after senators became frustrated at the industry's failure to cut emissions under previous, weaker air laws.

The EPA was also created under the National Environmental Policy Act about the same time as these additions were passed, which was important to help implement the programs listed above.

The Clean Air Act Amendments of 1977 required Prevention of Significant Deterioration (PSD) of air quality for areas attaining the NAAQS and added requirements for non-attainment areas.

The 1990 Clean Air Act added regulatory programs for control of acid deposition (acid rain) and stationary source operating permits. The amendments moved considerably beyond the original criteria pollutants, expanding the NESHAP program with a list of 189 hazardous air pollutants to be controlled within hundreds of source categories, according to a specific schedule. The NAAQS program was also expanded. Other new provisions covered stratospheric ozone protection, increased enforcement authority and expanded research programs.

History of the Clean Air Act

Introduction

President Lyndon B. Johnson signing the 1967 Clean
Air Act in the East Room of the White House, November 21, 1967.

The legal authority for federal programs regarding air pollution control is based on the 1990 Clean Air Act Amendments (1990 CAAA). These are the latest in a series of amendments made to the Clean Air Act (CAA), often referred to as "the Act." This legislation modified and extended federal legal authority provided by the earlier Clean Air Acts of 1963 and 1970.

The 1955 Air Pollution Control Act was the first federal legislation involving air pollution; it authorized $3 million per year to the U.S. Public Health Service for five years to fund federal level air pollution research, air pollution control research, and technical and training assistance to the states. Subsequently, the act was extended for four years in 1959 with funding levels at $5 million per year. The act was then amended in 1960 and 1962. Although the 1955 act brought the air pollution issue to the federal level, no federal regulations were formulated. Control and prevention of air pollution was instead delegated to state and local agencies.

The Clean Air Act of 1963 was the first federal legislation regarding air pollution control. It established a federal program within the U.S. Public Health Service and authorized research into techniques for monitoring and controlling air pollution. In 1967, the Air Quality Act was enacted in order to expand federal government activities. In accordance with this law, enforcement proceedings were initiated in areas subject to interstate air pollution transport. As part of these proceedings, the federal government for the first time conducted extensive ambient monitoring studies and stationary source inspections.

The Air Quality Act of 1967 also authorized expanded studies of air pollutant emission inventories, ambient monitoring techniques, and control techniques.

Clean Air Act of 1970

The *Clean Air Act of 1970* (1970 CAA) authorized the development of comprehensive federal and state regulations to limit emissions from both stationary (industrial) sources and mobile sources. Four major regulatory programs affecting stationary sources were initiated:

- the National Ambient Air Quality Standards [NAAQS (pronounced "knacks")],

- State Implementation Plans (SIPs),

- New Source Performance Standards (NSPS),

- and National Emission Standards for Hazardous Air Pollutants (NESHAPs).

Enforcement authority was substantially expanded. This very important legislation was adopted at approximately the same time as the *National Environmental Policy Act* . The U.S. Environmental Protection Agency was created by Executive Order of President Richard Nixon on December, 1970, and he appointed William Ruckelshaus as the first EPA Administrator.

Clean Air Act Amendments of 1977

Major amendments were added to the *Clean Air Act* in 1977 (1977 CAAA). The 1977 Amendments primarily concerned provisions for the Prevention of Significant Deterioration (PSD) of air quality in areas attaining the NAAQS. The 1977 CAAA also contained requirements pertaining to sources in non-attainment areas for NAAQS. A non-attainment area is a geographic area that does not meet one or more of the federal air quality standards. Both of these 1977 CAAA established major permit review requirements to ensure attainment and maintenance of the NAAQS.

Clean Air Act Amendments of 1990

Another set of major amendments to the Clean Air Act occurred in 1990 (1990 CAAA).

The 1990 CAAA substantially increased the authority and responsibility of the federal government. New regulatory programs were authorized for control of acid deposition (acid rain) and for the issuance of stationary source operating permits. The NESHAPs were incorporated into a greatly expanded program for controlling toxic air pollutants. The provisions for attainment and maintenance of NAAQS were substantially modified and expanded. Other revisions included provisions regarding stratospheric ozone protection, increased enforcement authority, and expanded research programs.

Milestones

Some of the principal milestones in the evolution of the Clean Air Act are as follows:

The Air Pollution Control Act of 1955

- First federal air pollution legislation
- Funded research on scope and sources of air pollution

Clean Air Act of 1963

- Authorized a national program to address air pollution
- Authorized research into techniques to minimize air pollution

Air Quality Act of 1967

- Authorized enforcement procedures involving interstate transport of pollutants
- Expanded research activities

Clean Air Act of 1970

- Established National Ambient Air Quality Standards
- Established requirements for State Implementation Plans to achieve them
- Establishment of New Source Performance Standards for new and modified stationary sources
- Establishment of National Emission Standards for Hazardous Air Pollutants
- Increased enforcement authority
- Authorized control of motor vehicle emissions

1977 Amendments to the Clean Air Act of 1970

- Authorized provisions related to prevention of significant deterioration
- Authorized provisions relating to non-attainment areas

1990 Amendments to the Clean Air Act of 1970

- Authorized programs for acid deposition control

- Authorized controls for 189 toxic pollutants, including those previously regulated by the national emission standards for hazardous air pollutants

- Established permit program requirements

- Expanded and modified provisions concerning National Ambient Air Quality Standards

- Expanded and modified enforcement authority

Regulations

Since the initial establishment of six mandated criteria pollutants (ozone, particulate matter, carbon monoxide, nitrogen oxides, sulfur dioxide, and lead), advancements in testing and monitoring have led to the discovery of many other significant air pollutants.

However, with the act in place and its many improvements, the U.S. has seen many pollutant levels and associated cases of health complications drop. According to the EPA, the 1990 Clean Air Act Amendments has prevented or will prevent:

	Year 2010 (cases prevented)	Year 2020 (cases prevented)
Adult Mortality - particles	160,000	230,000
Infant Mortality - particles	230	280
Mortality - ozone	4,300	71,000
Chronic Bronchitis	54,000	75,000
Heart Disease - Acute Myocardial Infarction	130,000	200,000
Asthma Exacerbation	1,700,000	2,400,000
Emergency Room Visits	86,000	120,000
School Loss Days	3,200,000	5,400,000
Lost Work Days	13,000,000	17,000,000

This chart shows the health benefits of the Clean Air Act programs that reduce levels of fine particles and ozone.

In 1997 EPA tightened the NAAQS regarding permissible levels of the ground-level ozone that make up smog and the fine airborne particulate matter that makes up soot. The decision came after months of public review of the proposed new standards, as well as long and fierce internal discussion within the Clinton administration, leading to the most divisive environmental debate of that decade. The new regulations were challenged in the courts by industry groups as a violation of the U.S. Constitution's non-delegation principle and eventually landed in the Supreme Court of the United States, whose 2001 unanimous ruling in *Whitman v. American Trucking Ass'ns, Inc.* largely upheld EPA's actions.

The Clean Air Act (CAA or Act) directs EPA to establish national ambient air quality standards (NAAQS) for pollutants at levels that will protect public health. EPA and American Lung Association promoted the 2011 Cross State Air Pollution Rule (CSAPR) to control ozone and fine particles. Aim was to cut emissions half from 2005 to 2014. It was claimed to prevent each year 400,000 asthma cases and save ca 2m work and schooldays lost by respiratory illness. Some states (e.g. Texas), cities and power companies sued the case (EPA v EME Homer City Generation). The appeals-court judges decided by two to one that the rule is too strict. Based on appeals the power companies were allowed to continue thousands of persons respiratory illnesses prolonged time in the USA. According to the Economist (2013) the Supreme Court decision may affect how the EPA regulates other pollutants, including the greenhouse gases.

Roles of the Federal Government and States

Although the 1990 Clean Air Act is a federal law covering the entire country, the states do much of the work to carry out the Act. The EPA has allowed the individual states to elect responsibility for compliance with and regulation of the CAA within their own borders in exchange for funding. For example, a state air pollution agency holds a hearing on a permit application by a power or chemical plant or fines a company for violating air pollution limits. However, election is not mandatory and in some cases states have chosen to not accept responsibility for enforcement of the act and force the EPA to assume those duties.

In order to take over compliance with the CAA the states must write and submit a state implementation plan (SIP) to the EPA for approval. A state implementation plan is a collection of the regulations a state will use to clean up polluted areas. The states are obligated to notify the public of these plans, through hearings that offer opportunities to comment, in the development of each state implementation plan. The SIP becomes the state's legal guide for local enforcement of the CAA. For example, Rhode Island law requires compliance with the Federal CAA through the SIP. The SIP delegates permitting and enforcement responsibility to the state Department of Environmental Management (RI-DEM).

The federal law recognizes that states should lead in carrying out the Clean Air Act, because pollution control problems often require special understanding of local industries, geography, housing patterns, etc. However, states are not allowed to have weaker pollution controls than the national minimum criteria set by EPA. EPA must approve each SIP, and if a SIP isn't acceptable, EPA can take over CAA enforcement in that state.

The United States government, through the EPA, assists the states by providing scientific research, expert studies, engineering designs, and money to support clean air programs.

Metropolitan planning organizations must approve all federally funded transportation projects in a given urban area. If the MPO's plans do not, Federal Highway Administration and the Federal Transit Administration have the authority to withhold funds if the plans do not conform with federal requirements, including air quality standards. In 2010, the EPA directly fined the San Joaquin Valley Air Pollution Control District $29 million for failure to meet ozone standards, resulting in fees for county drivers and businesses. This was the results of a federal appeals court case that required the EPA to continue enforce older, stronger standards, and spurred debate in Congress over amending the Act.

State Programs

Many states, or concerned citizens of the state, have established their own programs to help promote pollution clean-up strategies.

For example,(in alphabetical order by state)

- California - California's Clean Air Project - designed to create a smoke-free gaming atmosphere in tribal casinos

- Georgia - The Clean Air Campaign

- Illinois - Illinois Citizens for Clean Air and Water - coalition of farmers and other citizens to reduce harmful effects of large-scale livestock production methods

- New York - Clean Air NY

- Oklahoma - "Breathe Easy" - Oklahoma Statutes on Smoking in Public Places and Indoor Workplaces (Effective November 1, 2010)

- Texas - Drive Clean Across Texas

- Virginia - Virginia Clean Cities, Inc.

Interstate air Pollution

Air pollution often travels from its source in one state to another state. In many metropolitan areas, people live in one state and work or shop in another; air pollution from cars and trucks may spread throughout the interstate area. The 1990 Clean Air Act provides for interstate commissions on air pollution control, which are to develop regional strategies for cleaning up air pollution. The 1990 amendments include other provisions to reduce interstate air pollution.

The Acid Rain Program, created under Title IV of the Act, authorizes emissions trading to reduce the overall cost of controlling emissions of sulfur dioxide.

Leak Detection and Repair

The Act requires industrial facilities to implement a Leak Detection and Repair (LDAR) program to monitor and audit a facility's fugitive emissions of volatile organic compounds (VOC). The program is intended to identify and repair components such as valves, pumps, compressors, flanges, connectors and other components that may be leaking. These components are the main source of the fugitive VOC emissions.

Testing is done manually using a portable vapor analyzer that read in parts per million (ppm). Monitoring frequency, and the leak threshold, is determined by various factors such as the type of component being tested and the chemical running through the line. Moving components such as pumps and agitators are monitored more frequently than non-moving components such as flanges and screwed connectors. The regulations require that when a leak is detected the component be repaired within a set amount of days. Most facilities get 5 days for an initial repair attempt with no more than 15 days for a complete repair. Allowances for delaying the repairs beyond the allowed time are made for some components where repairing the component requires shutting process equipment down.

Application to Greenhouse Gas Emissions

EPA began regulating greenhouse gases (GHGs) from mobile and stationary sources of air pollution under the Clean Air Act for the first time on January 2, 2011. Standards for mobile sources have been established pursuant to Section 202 of the CAA, and GHGs from stationary sources are controlled under the authority of Part C of Title I of the Act.

Below is a table for the sources of greenhouse gases, taken from data in 2008. Of all greenhouse gases, about 76 percent of the sources are manageable under the CAA, marked with an asterisk (*). All others are regulated independently, if at all.

Source	Percentage
Electric Generation*	34%
Industry*	15%
Large Non-Agricultural Methane Sources*	5%
Light-, Medium-, and Heavy-Duty Vehicles*	22%
Other Transport	7%
Commercial and Residential Heating	7%
Agriculture	7%
HFCs	2%
Other	1%

References

- Jacobson, Mark Z. (April 2012). Air Pollution and Global Warming History, Science, and Solutions (Google Books) (2nd ed.). Cambridge University Press. pp. 175, 176. ISBN 9781107691155.

- Shekhtman, Lonnie. "Beijing smog: What makes some cities cleaner than others?". Christian Science Monitor. ISSN 0882-7729. Retrieved 2015-12-22.

- Enesta Jones (03/01/2011). "EPA Report Underscores Clean Air Act's Successful Public Health Protections/Landmark law saved 160,000 lives in 2010 alone". EPA.gov. Retrieved 22 March 2012.

- "Clean Air Act: Title I - Air Pollution Prevention and Control". U.S. Environmental Protection Agency (EPA). Retrieved 29 April 2012.

- Trendowski, John. "Sustainability Trends — Reducing Emissions at Airports" (PDF). Airport Magazine. Retrieved 22 April 2012.

- McCarthy, James. "Clean Air Act: A Summary of the Act and its Major Requirements" (PDF). CRS Report for Congress. Retrieved 23 April 2012.

Permissions

All chapters in this book are published with permission under the Creative Commons Attribution Share Alike License or equivalent. Every chapter published in this book has been scrutinized by our experts. Their significance has been extensively debated. The topics covered herein carry significant information for a comprehensive understanding. They may even be implemented as practical applications or may be referred to as a beginning point for further studies.

We would like to thank the editorial team for lending their expertise to make the book truly unique. They have played a crucial role in the development of this book. Without their invaluable contributions this book wouldn't have been possible. They have made vital efforts to compile up to date information on the varied aspects of this subject to make this book a valuable addition to the collection of many professionals and students.

This book was conceptualized with the vision of imparting up-to-date and integrated information in this field. To ensure the same, a matchless editorial board was set up. Every individual on the board went through rigorous rounds of assessment to prove their worth. After which they invested a large part of their time researching and compiling the most relevant data for our readers.

The editorial board has been involved in producing this book since its inception. They have spent rigorous hours researching and exploring the diverse topics which have resulted in the successful publishing of this book. They have passed on their knowledge of decades through this book. To expedite this challenging task, the publisher supported the team at every step. A small team of assistant editors was also appointed to further simplify the editing procedure and attain best results for the readers.

Apart from the editorial board, the designing team has also invested a significant amount of their time in understanding the subject and creating the most relevant covers. They scrutinized every image to scout for the most suitable representation of the subject and create an appropriate cover for the book.

The publishing team has been an ardent support to the editorial, designing and production team. Their endless efforts to recruit the best for this project, has resulted in the accomplishment of this book. They are a veteran in the field of academics and their pool of knowledge is as vast as their experience in printing. Their expertise and guidance has proved useful at every step. Their uncompromising quality standards have made this book an exceptional effort. Their encouragement from time to time has been an inspiration for everyone.

The publisher and the editorial board hope that this book will prove to be a valuable piece of knowledge for students, practitioners and scholars across the globe.

Index

www.ingramcontent.com/pod-product-compliance
Lightning Source LLC
Chambersburg PA
CBHW080239230326
41458CB00096B/2712